Lecture Notes in
Mathematics

Edited by A. Dold and B. Eckmann

983

Nonstandard Analysis-
Recent Developments

Edited by A. E. Hurd

Springer-Verlag
Berlin Heidelberg New York Tokyo 1983

Editor

Albert Emerson Hurd
Department of Mathematics, University of Victoria
Victoria, British Columbia, Canada V8W2Y2

AMS Subject Classifications (1980): 46A05, 82A05, 90A99, 34C99, 60G05, 03H05, 03H10

ISBN 3-540-12279-6 Springer-Verlag Berlin Heidelberg New York Tokyo
ISBN 0-387-12279-6 Springer-Verlag New York Heidelberg Berlin Tokyo

Printed in Germany

Printing and binding: Beltz Offsetdruck, Hemsbach/Bergstr.
2146/3140-543210

PREFACE

The papers in this volume were collected together to record the continuing advances being made in the applications of nonstandard analysis. Some of the papers are elaborations of work presented to the Second Victoria Symposium on Nonstandard Analysis held at the University of Victoria, Victoria, Canada, in 1980.

I would like to thank Professors Robert Anderson, H. Jerome Keisler, Peter Loeb and W.A.J. Luxemburg for their help in editing the volume.

A.E. HURD

Nov. 10, 1982

CONTENTS

Explicit Solutions of Partial Differential Equations

M. Berger and A. Sloan

School of Mathematics
Georgia Institute of Technology
Atlanta, Georgia 30332

I. Introduction.

The classical method of characteristics for the solution of the Cauchy problem for a first order evolution equation

$$u_t = a(x)u_x \qquad u(x,0) = f(x)$$

consists in writing $u(x,t) = f(\eta_x(t))$ where $\eta_x(t)$ is the solution of a corresponding characteristic equation $d\eta_x(t) = a(\eta_x(t))dt$. The subscript x is a parameter indicating the initial condition $\eta_x(0) = x$.

The analogous solution of the heat equation

$$u_t = \frac{1}{2} a^2(x)u_{xx} , \ u(x,0) = f(x)$$

requires the introduction of the differential of Brownian motion, which we denote by $(dt)^{1/2}$. The associated characteristic is no longer defined as the solution to an ordinary differential equation but rather as the solution of the stochastic differential equation

$$d\eta_x = a(\eta_x)(dt)^{1/2}, \ \eta_x(0) = x .$$

The solution may then be written as

$$u(x,t) = E(f(\eta_x(t)))$$

where E is an expectation operator.

In [Berger and Sloan, 1] differentials $(dt)^r$, for rational r in (0,1] were informally introduced, together with a calculus sufficient to permit the explicit solution of evolution equations

$$u_t = Qu, \ u(x,0) = f(x)$$

for a wide variety of partial differential operators Q having arbitrary order. The method was made rigorous for constant coefficient equations in [Berger and Sloan, 2].

However, the generalized characteristic η_x were not defined pointwise in time, t. The existence of a realization of the structure in terms of characteristics $\eta_x(t)$ defined for $t \geq 0$ is important in generalizing the applications to boundary-initial value problems where the existence of hitting times should prove useful.

In this paper, the authors show that the solutions to generalized characteristic equations such as

$$d\eta_x = a_1(dt)^{1/n} + a_2(dt)^{2/n} + \cdots + a_n(dt)^1$$
$$\eta_x(0) = x$$

can, for constant $a_1, \cdots, a_n$, be realized for each x in $\mathbb{R}$ and $t > 0$ as a function, $\eta_x(t)$, from path space $\Omega = \{w: [0,\infty) \to \mathbb{R}\}$ into the extended complex numbers $*\mathbb{C}$. A generalized expectation operator E is introduced to map functions of these characteristic into finite elements of $*\mathbb{C}$. Each constant coefficient $Q = \sum b_i \partial^i/\partial x^i$ is associated with a generalized characteristic η_x in such a way that the solution to

$$u_t = Qu \qquad u(x,0) = f(x)$$

may be represented as $u(x,t) = {}^\circ Ef(\eta_x(t))$.

Nonstandard analysis appears twice in this work. First, the Gaussian kernel which arises from the heat equation must be replaced by the Fourier transform in λ of $e^{(i\lambda)^q t/q!}$. This is done via an infinite cutoff. The authors expect this use to eventually disappear by considering only cases when q is an odd multiple of 2 through an embedding result. The second use appears in the definition of integrals such as

$$\int_0^t f(x(s))(ds)^r$$

for r rational in (0,1]. Although standard weak definitions are possible, see [Hochberg, 3] nonstandard analysis permits such integrals to be viewed as functions on path space, for each t.

II. Generalized Expectations.

Fix α to be a positive infinite integer and fix q to be positive finite integer. Then define

$$p(x,t) = (2\pi)^{-1}\int_{-\alpha}^{\alpha} e^{i\lambda x} e^{(i\lambda)^q t/q!} d\lambda \ . \tag{1}$$

p is to induce an expectation type operator on a space Ω of paths as follows:

$\Omega = \{x: [0,\infty) \to \mathbb{R}^1\}$. A tame function $g = (f,\bar{t})$ on Ω consists of a complex valued function of k real variables and a set $\bar{t} = (t_1,\cdots,t_k)$ of positive times $t_1 < t_2 < \cdots < t_k$. We view g as a function on Ω according to the rule

$$g(x) = f(x(t_1),\cdots,x(t_k)) . \tag{2}$$

Here $k = 1,2,3,\cdots$. We then restrict the set of tame functions further by requiring f to be in a convenient vector space of functions for which the generalized expectation E of g, defined by

$$E(g) = \int_{*\mathbb{R}^k} f(\bar{x}) \prod_{i=1}^{k} p(\Delta x_i , \Delta t_i) d\bar{x} \tag{3}$$

exists in some reasonable sense. In (3) $\bar{x} = (x_1,\cdots,x_k)$, $\Delta x_i = x_i - x_{i-1}$, $\Delta t_i = t_i - t_{i-1}$, $t_0 = x_0 = 0$ and $d\bar{x} = dx_1 \cdots dx_k$. We also write

$$E(g) = E(f(x(t_1),\cdots,x(t_k))) .$$

For example, let $f(x) = e^{-iax}$. For fixed t, $p(\cdot,t)$ is the inverse Fourier transform of $h(\cdot,t)$, where $h(\lambda,t) = \chi_\alpha(\lambda)e^{(i\lambda)^q t/q!}$, χ_α being the characteristic function of $(-\alpha,\alpha)$. Since h is differentiable, except at $\lambda = \pm\alpha$ where h has simple jump discontinuities we compute

$$E(e^{-iax(t)}) = \begin{cases} 0 & |a| > \alpha \\ e^{(ia)^q t/q!} & |a| < \alpha \end{cases} \tag{4}$$

Our goal is to extend E to $(f,\bar{t})$ where $f \in Z$, the space of entire functions of exponential type.

Let $C^\omega(\mathbb{R}^k)$ be the complex valued analytic function on $\mathbb{R}^k$. For each $a > 1$ let

$$Z_a(k) = \{f \in C^\omega(\mathbb{R}^k) : \sup_\mu a^{-|\mu|} |\partial^\mu f(0)|\} \tag{5}$$

where the sup is over all $\mu = (\mu_1,\cdots,\mu_k)$, $\mu_i = 0,1,2,\cdots$, $|\mu| = \mu_1+\cdots+\mu_k$ and

$$\partial^\mu f = \partial^{|\mu|} f/(\partial x_1^{\mu_1})\cdots(\partial x_k^{\mu_k})$$

$f \in Z_a(k)$ if and only if there is a constant M so the coefficients, c_μ, in the power series expansion for $f = \sum c_\mu x^\mu$ satisfy

$$|c_\mu| \le M\, a^{|\mu|}/|\mu|! .$$

If $|f|_a$ denotes the inf over all such M then $|f|_a = \sup_\mu |\partial^\mu f(0)| a^{-|\mu|}$ and $|\ |_a$ defines a norm $Z_a(k)$ which turns $Z_a(k)$ into a Banach space. Norm convergence on $Z_a(k)$ implies pointwise convergence of all derivatives uniformly on compact sets.

Let $Z(k) = \bigcup_{a\geq 1} Z_a(k)$ and $Z = \bigcup_k Z_k$.

We now extend E to $Z \cap L^2$ according to the formula

$$E(f(x(\bar{t}))) = (2\pi)^{-k} \int_{*\mathbb{R}^k} \hat{f}(\bar{\omega}) E(e^{i\omega\cdot x(\bar{t})}) d\bar{\omega} \tag{6}$$

where $x(\bar{t}) = (x(t_1),\cdots,x(t_k))$, $\bar{\omega} = (\omega_1,\cdots,\omega_k)$, $\hat{f}$ is the Fourier transform of f and $d\bar{\omega} = d\omega_1\cdots d\omega_k$. Equation (6) makes sense because $\hat{f} \in L^2$ has compact support whenever $f \in Z \cap L^2$ by the Paley-Wiener theorem and $E(e^{i\bar{\omega}\cdot x(\bar{t})})$ is a continuous function of $\bar{t} = (t_1,\cdots,t_k)$.

Two observations may be made at this point: (i) definitions (3) and (6) coincide for f in $Z \cap L^2$; and (ii) if f is standard and if each t_i is standard then so is $E\, f(x(\bar{t}))$.

Before proceeding with the further extension of E we wish to note several properties of a probabilistic nature. $E(1) = 1$ follows from (4) even though in general there is no underlying probability space. E is also linear. Viewing E as a map from functions on the "<u>process</u>" x, we say our process has <u>independent increments</u> because

$$E[f(\Delta_k x)\cdot g(x(\bar{t}))] = E[f(\Delta_k x)]\cdot E[g(x(\bar{t}))] \tag{7}$$

whenever $\Delta_k x = x(t_k) - x(t_{k-1})$ and $x(\bar{t}) = (x(t_1),\cdots,x(t_{k-1}))$ where $t_1 < t_2 < \cdots < t_k$. In order to verify (7) first observe that if $\omega \in Z \cap L^2(\mathbb{R}^\ell)$ then also $\omega_0 \in Z \cap L^2(\mathbb{R}^\ell)$ where

$$\omega_0(y_1,\cdots,y_\ell) = \omega(y_1, y_1+y_2,\cdots,y_1+\cdots+y_\ell) .$$

We then write

$$\begin{aligned} E[f(\Delta_k x)\cdot g(x(\bar{t}))] &= \int f(\Delta x_k) g(\bar{x}) \prod_1^k p(\Delta x_i, \Delta t_i) d\bar{x}\, dx_k \\ &= \int f(\Delta x_k) g_0(\Delta x_1,\cdots,\Delta x_{k-1}) \prod_1^k p(\Delta x_i,\Delta t_i) d\bar{x}\, dx_k \\ &= \int f(z) p(z,\Delta t_k) dz \int g_0(\Delta x_1,\cdots,\Delta x_{k-1}) \prod_1^{k-1} p(\Delta x_i,\Delta t_i) d\bar{x} \\ &= \int f(z-y) p(z-y,\Delta t_k) p(y,t_{k-1}) dz\, dy \cdot E[g(x(\bar{t}))] \end{aligned}$$

because $1 = E(1) = \int p(x,s)dx$. This proves (7). Our proof also revealed the fact that the process is <u>stationary</u> in the sense that

$$E[f(x(t+s)-x(s))] = E[f(x(t))] \,. \tag{8}$$

In addition to the probabilistic identities the further extension of E will require approximation in Z. Z(k) is a union of topological spaces and convergence in Z(k) may be defined according to the rule

$(f_n \to f)$ if and only if

(there is an a with f_n, f in $Z_a(k)$ for all m sufficiently large and $f_n \to f$ in $Z_a(k)$).

Although polynomials are not dense in any Z_a they are dense in Z in the sense that if $f = \sum c_\mu x^\mu$ belongs to $Z_a(k)$ then $f_m = \sum_{|\mu|\le m} c_\mu x^\mu \to f$ in $Z_b(k)$, for all $b > a$. We also note that $|f_m|_b \le |f|_b$ and $|x^\mu|_b = |\mu|!\,b^{-|\mu|}$.

In a similar sense, $Z \cap L^2$ is dense in Z for let $g^\circ(x) = \sin x/x$ and $g(x) = 2(1-\cos x)/x$. Then $g^n \in L^2 \cap Z(1)$ for all $n = 0,1,2,\cdots$. Let $\varepsilon_0 = \varepsilon$ and $\varepsilon_k = \varepsilon/(2^k \cdot k!(k+1)^{k+1})$ for any $\varepsilon > 0$. In particular if $c > 0$ and $f = \sum c_\mu x^\mu \in Z_a(\ell)$ are given choose for each $m = 1,2,\cdots$ $\varepsilon = c/2(m+1)$ and define

$$f_m(x) = \sum_{|\mu|\le m} c_\mu \frac{1}{\varepsilon_\mu} g^\mu(\varepsilon_\mu x)$$

where for $\mu = (\mu_1,\cdots,\mu_\ell)$, $\varepsilon_\mu = \varepsilon_{\mu_1}\cdots\varepsilon_{\mu_\ell}$ and $g^\mu(\varepsilon_\mu x) = g^{\mu_1}(\varepsilon_{\mu_1} x_1)\cdots g^{\mu_\ell}(\varepsilon_{\mu_\ell} x_\ell)$. Then for all m sufficiently large and $b > a$ we have $|f-f_m|_b < c$.

In [2] we showed that for each $a > 1$ there is an M such that if $f \in L^2 \cap Z_a(1)$ and if $b > M$

$$|E\, f(x(t))|_b \le |f|_a \,. \tag{9}$$

Consequently we may extend E to all f in Z(1) in such a way that (1) holds by $L^2 \cap Z$ approximation.

Next let $f(x_1,\cdots,x_k)$ be a polynomial and define $f_1(x_1,\cdots,x_k) = f(x_1,\cdots,x_{k-1}, x_k+x_{k+1})$. Note that $|f_1|_{4b} \le |f|_b$ for all $b \ge 1$. We may now define $E(f(x(t_1),\cdots,x(t_k)) = E(f_1(x(t_1),\cdots,x(t_{k-1}),x(t_k)-x(t_{k-1}))$ inductively on k using the fact that we just defined E on such f for $k = 1$ and using linearity and the independent increment property. Our immediate aim is to show that for each a and k there is an M such that for all $b > M$ and for all polynomials f in k variables

$$|E[f(x(t_1),\cdots,x(t_k))]|_b \le |f|_a \tag{10}$$

where we consider $E[f(x(t_1),\cdots,x(t_k))]$ to be a function of $t_1,\cdots,t_k$. In [Berger and Sloan, 2] we saw that for any polynomial f of one variable, $Ef(x(t))$ is a polynomial in t. Our inductive definition consequently insures that $E[f(x(\bar{t}))]$ is a polynomial in $\bar{t} = (t_1,\cdots,t_k)$, for all polynomials $f = \sum c_\mu x^\mu$. In this case we also have

$$|f|_b = \max_\mu |\mu|!c_\mu b^{-|\mu|}.$$

We now prove (10) by induction on k. From (9) we see that (10) is true for $k = 1$. Suppose in addition (10) is true for $k = 1,2,\cdots,\ell$. Let $\bar{x} = (x_1,\cdots,x_\ell)$, $\bar{t} = (t_1,\cdots,t_\ell)$. Let $f(\bar{x},y)$ be a polynomial in $\ell+1$ variables. Write

$$f_1(\bar{x},y) = \sum p_n(\bar{x})y^n$$

where each p_n is a polynomial. Let $q_n(\bar{t}) = E[p_n(x(\bar{t}))]$ and $d_n(s) = E(x(s)^n)$. Set $q(\bar{t},s) = \sum q_n(\bar{t})d_n(s)$.

Let $a > 1$ be given. Choose M_ℓ according to our inductive hypothesis so that for $b > M_\ell$, $|q_n|_b \le |p_n|_{4a}$ and $M_\ell > M_{\ell-1} > \cdots M_1$. Let $M_{\ell+1} = 8M_\ell$ and now choose $b > M_{\ell+1}$. Then

$$\begin{aligned}
|E[f(x(\bar{t}),x(s)]|_b &= |E[f_1(x(\bar{t}),x(s)-x(t_\ell))]|_b \\
&= |\sum E[p_n(x(\bar{t}))]\cdot E[(x(s)-x(t_\ell))^n]|_b \\
&= |\sum q_n(\bar{t})d_n(s-t_\ell)|_b \\
&\le |q(\bar{t},s)|_{b/4} \\
&= \max_n |q_n(\bar{t})E[x(s)^n]|_{b/4} \\
&= |E[p_m(t)]E[x(s)^n]|_{b/4} \\
&\le |E[p_m(\bar{t})]|_{b/8}|E[x(s)]^m]|_{b/8} \\
&\le |p_m|_{4a}\ m!(4a)^{-m} \\
&\le |p_m(\bar{x})y^m|_{4a} \\
&\le |f_1(\bar{x},y)|_{4a} \\
&\le |f(\bar{x},y)|_a .
\end{aligned}$$

The bound (10) is now valid for all polynomials. Since polynomials are dense in Z, E extends to $f(x(\bar{t}))$ for all f in Z in such a way so as to preserve the validity of (10). Furthermore this extension is linear, takes on standard values whenever f and $\bar{t}$ are standard. Finally, the extension has independent increments and is stationary in the sense that (7) and (8) hold for all f in Z.

II. Stochastic Calculus.

In this section we extend E to certain functions $f(x(\bar{t}))$ where $\bar{t} = (t_1, \cdots, t_k)$ and k is infinite. Since the constant M occuring in (10) depends on k it is not apriori evident that this extension can be done in a near standard way.

Fix m to be a positive infinite integer. For each finite $t > 0$ set $\ell = \ell(t) = [mt]$, the greatest integer less than mt. For $i = 0,1,\cdots,\ell$ set $x_i = x(k/m)$ and $x_{\ell+1} = x(t)$. Thus each $x_i : \Omega \to {}^*\mathbb{R}$. For given functions f and g define

$$S_m(f\cdot g(dx))\Big|_0^t = \sum_{i=1}^{\ell+1} f(x_{i-1})g(\Delta x_i)$$

as a map from Ω into ${}^*\mathbb{R}$. We are interested in the case $g(x) = x^j$ for finite positive integers j and f in Z. We shall denote this map, for $g(x) = x^j$, by

$$\int_0^t f(x(s))(ds)^{j/q} .$$

Here q remains the fundamental constant introduced at the start of I.

We extend E to these generalized stochastic integrals by linearity as

$$E(\int_0^t f(x(s))(ds)^{j/q}) = \sum_{i=1}^{\ell+1} E[f(x_i)g(\Delta x_i)] ,$$

an element of ${}^*\mathbb{R}$. In [Berger and Sloan, 2] we calculated

$$E(x(t)^{rq+p}) = \begin{cases} 0 & \text{if } p = 1,2,\cdots,q-1 \\ (rq)!t^t/(q!)^r r! & \text{if } p = 0 , \end{cases}$$

for all non-negative integers r. Consequently,

$$E(\int_0^t f(x(s))(ds)^{j/q}) = 0 \tag{11}$$

whenever $j = rq + p$ and $p = 1,2,\cdots,q-1$, $r = 0,1,2,\cdots$. If $p = 0$ so that $j = rq$ then

$$E[\int_0^t f(x(s))(ds)^{j/q}] = \sum_1^{\ell} E[f(x(t_{i-1}))] \frac{(rq)!}{(q!)^r r!}\cdot\left(\frac{1}{m}\right)^r + Ef(x(t)) \frac{(rq)!}{(q!)^r r!}\left(t-\left(\frac{\ell}{m}\right)\right)^r ,$$

where we have used the independent increment property. Since $E(f(x(s)))$ is an analytic function of s we find

$$E(\int_0^t f(x(s))(ds)^r) \underset{\sim}{\approx} m^{1-r}\int_0^t E[f(x(s))]ds .$$

Thus

$$(12)\qquad E\Big(\int_0^t f(x(s))(ds)^r\Big) \approx \begin{cases} 0 & r = 2,3,\cdots \\ \int_0^t E[f(x(s))]ds & r = 1 \end{cases}$$

Stochastic integration produces new processes: $t_j(r) = \int_0^r (ds)^j$ for $j = \frac{1}{q}, \frac{2}{q}, \cdots, 1$ so that $t_j(r)\colon \Omega \to {}^*\mathbb{R}$. We now try to extend E to functions of these processes, beginning with polynomials. We start with some calculations which relate such polynomial functions to stochastic integrals of other polynomial functions having lower order.

<u>Example</u>: If $f(x) = x^n$ then

$$f(t_j(t)) = \sum_{r=1}^{n} \frac{1}{r!} \int_0^t \left(\frac{d^r f}{dx^r}\right)(t_j(s))(ds)^{rj} .$$

We shall verify this by induction on n. It is true by definition in case n = 1. Suppose it is true for $n = 1,2,\cdots,w-1$. Then

$$(t_j(t))^w = \Big(\sum_{i=1}^{\ell+1}(\Delta_i x)^{jq}\Big)^{w-1}\Big(\sum_{k=1}^{\ell+1}(\Delta_k x)^{jq}\Big)$$

$$= \Big(\sum_{r=1}^{w-1} \frac{1}{r!}(w-1)\cdots(w-r)\Big(\sum_{i=1}^{\ell+1}\Big(\sum_{h=1}^{i-1}(\Delta_h x)^{jq}\Big)^{w-1-r}(\Delta_i x)^{rjq}\Big)\Big)\Big(\sum_{k=1}^{\ell+1}(\Delta_k x)^{jq}\Big)$$

$$= \sum_{r=1}^{w-1}\frac{1}{r!}(w-1)\cdots(w-r)\Big(\sum_{i=1}^{\ell+1}\Big(\sum_{h=1}^{i-1}(\Delta_h x)^{jq}\Big)^{w-1-r}(\Delta_i x)^{(r+1)jq}$$

$$+ \sum_{r=1}^{w-1}\frac{1}{r!}(w-1)\cdots(w-r)\Big(\sum_{i=1}^{\ell+1}\Big(\sum_{h-1}^{i-1}(\Delta_h x)^{jq}\Big)^{w-r}(\Delta_i x)^{rjq}\Big)$$

$$+ \sum_{r=1}^{w-1}\frac{1}{r!}(w-1)\cdots(w-r)\Big(\sum_{k=1}^{\ell+1}\Big(\sum_{i=1}^{k-1}\Big(\sum_{h=1}^{i-1}(\Delta_h x)^{jq}\Big)^{w-1-r}(\Delta_i x)^{rjq}(\Delta_k x)^{jq}$$

$$= \sum_{r=1}^{w-1}\frac{1}{r!}(w-1)\cdots(w-r)\Big(1+\frac{(w-(r+1))}{r+1}\Big)\int_0^t (t_j(s))^{w-(r+1)}(ds)^{((r+1)j)}$$

$$+ (w-1)\sum_{i=1}^{\ell+1}\Big(\sum_{h=1}^{i-1}(\Delta_h x)^{jq}\Big)^{w-1}(\Delta_i x)^{jq}$$

$$+ \sum_{k=1}^{\ell+1}\Big(\sum_{r=1}^{w-1}\frac{1}{r!}(w-1)\cdots(w-s)\Big(\sum_{i=1}^{k-1}\Big(\sum_{h=1}^{i-1}(\Delta_h x)^{jq}\Big)^{w-1-r}(\Delta_i x)^{rjq}\Big)(\Delta_k x)^{jq}\Big)$$

$$= \sum_{r=2}^{w}\frac{1}{r!}\int_0^t \frac{d^r f}{dx^r}(t_j(s))(ds)^{rj/q} + ((w-1)+1)\int_0^t \frac{df}{dx}(t_j(s))(ds)^{j/q} .$$

This example illustrates a generalized Taylor's formula. In order to develop it further we will need stochastic integrals for functions of several variables sucl as

$$\eta_i = x_i + \sum_{j=1}^{q} a_{ij} t^{j/q}$$

where for each $i = 1,\cdots,n$, x_i is real and a_{ij} is complex. We denote x_i by $\eta_i(0)$. For any polynomial h we define, for $\bar{\eta} = (\eta_1,\cdots,\eta_n)$

$$\int_0^t h(\bar{\eta}(s))(ds)^{r/q}$$

as the map from Ω into $^*\mathbb{C}$ which takes x into

$$\sum_{i=1}^{\ell+1} h(\bar{\eta}(x(\frac{i-1}{m})))(\Delta x_i)^r = \sum_{i=1}^{\ell+1} h(x_1 + \sum_{j=1}^{q} a_{ij}(\sum_{r_1=1}^{i-1}(\Delta x_{r_i})^j),\cdots,x_n$$

$$+ \sum_{j=1}^{q} a_{nj}(\sum_{r_n=1}^{i-1}(\Delta x_{r_n})^j))(\Delta x_i)^r$$

A calculation similar to the one in the previous example shows

$$\text{(13)} \qquad h(\bar{\eta}(t)) = h(\bar{\eta}(0)) + \sum_r \int_0^t (D_r h(\bar{\eta}(s)))(ds)^r$$

where the sum is over all $r = 1/q, 2/q, \cdots,$

$$\text{(14)} \qquad D_r h(\bar{\eta}(s)) = h_{r,\bar{\eta}}(\bar{\eta}(s)) \quad \text{and}$$

$$\text{(15)} \qquad h_{r,\bar{\eta}}(\bar{x}) = \sum_{k=1}^{\infty} \frac{1}{k!} \sum_{\mu\epsilon(n)^k} \Big(\sum_{\substack{j\epsilon(q)^k \\ |j|=r\cdot q}} (a_{\mu_1 j_1} \cdots a_{\mu_k j_k}) \Big) \partial_\mu h(\bar{x})$$

In (15), $(n)^k = \{(\mu_1,\cdots,\mu_k): \mu_i \epsilon \{1,2,\cdots,n\}\}$, $|\mu| = \mu_1+\cdots+\mu_k$ and $\partial_\mu h = \dfrac{\partial^k h}{\partial x_{\mu_1} \cdots \partial x_{\mu_k}}$. Since h is a polynomial all sums in (13) and (15) are finite.

Informally, if we define

$$d\eta_i = \sum_{i=1}^{q} a_{ij}(ds)^{j/q}$$

and extend the rule $(ds)^a(ds)^b = (ds)^{a+b}$ multilinearly to define $d_\mu \bar{\eta} = (d\eta_{\mu_1})\cdots(d\eta_{\mu_k})$ we may recover (13) from the differential rule

(16) $$dh(\bar{\eta}) = \sum_{k=1}^{\infty} \frac{1}{k!} \sum_{\mu \in (h)^k} \partial_\mu h(\bar{\eta}) d_\mu \bar{\eta}$$

by combining all terms of similar exponent in (ds) and integrating.

Linearity naturally defines E on $h(\bar{\eta}(t))$ and on $\int_0^t h(\bar{\eta}(s))(ds)^{r/q}$ as elements of $^*\mathbb{C}$. The independent increment property shows that

$$E(\int_0^t h(\bar{\eta}(s))(ds)^{r/q}) = 0$$

if r/q is not a positive integer.

Our next goal is to demonstrate that for any fixed standard $T > 0$, the function $s \to Eh(\bar{\eta}(s))$ is uniformly continuous on $^*[0,T]$ in the standard sense. That is, given any standard $\varepsilon > 0$ there is a standard $\delta > 0$ such that whenever s,t are in $^*[0,T]$ with $|s-t| < \delta$ then $|Eh(\bar{\eta}(s)) - Eh(\bar{\eta}(t))| < \varepsilon$.

One immediate consequence of this demonstration will be the conclusion that, since $E(h(\bar{\eta}(0)) = h(\bar{\eta}(0))$ is standard, there is a standard $K > 0$ such that $|Eh(\bar{\eta}(s))| < K$ for all s in $^*[0,T]$. Each $Eh(\bar{\eta}(s))$ is then near standard for s finite and if we denote its standard part by $^\circ Eh(\bar{\eta}(s))$ then $s \to {}^\circ Eh(\bar{\eta}(s))$ is continuous on $[0,\infty)$.

The proof of the standard uniform continuity of $Eh(\bar{\eta}(1))$ is by induction on the degree of the standard polynomial h. If h has degree zero then $Eh(\bar{\eta}(s)) = Eh(\bar{\eta}(0))$ and the result is true. Now assume this continuity for all polynomials of degree less than that of h and, in particular, for $h_{r,\bar{\eta}}$ where $D_r h(\bar{\eta}(s)) = h_{r,\bar{\eta}}(\bar{\eta}(s))$. The finite uniform bound of $E(D_r h(\bar{\eta}(s)))$ for s in $^*[0,T]$ shows that

(17) $$Eh(\bar{\eta}(t)) \approx Eh(\bar{\eta}(0)) + E(\int_0^t D_1 h(\bar{\eta}(s))(ds)^1)$$

Now suppose $t < s$. Then for any standard $\varepsilon > 0$ (17) implies that

$$|Eh(\bar{\eta}(t)) - Eh(\bar{\eta}(s))| < \varepsilon + \sum_{t \le \frac{i}{m} \le s} ED_1 h(\bar{\eta}(\tfrac{i}{m})) \cdot \frac{1}{m} .$$

For $s - t$ sufficiently small, the standard uniform continuity of $ED_1 h(\bar{\eta}(\cdot))$ implies that

$$|Eh(\bar{\eta}(t)) - Eh(\bar{\eta}(s))| < \varepsilon + (s-t)\varepsilon < (T+1)\varepsilon$$

and so $Eh(\bar{\eta}(\cdot))$ is uniformly continuous in the standard sense on $^*[0,T]$. This continuity also shows

$$(18) \qquad E\left(\int_0^t Dh(\bar{\eta}(s))(ds)^1\right) \approx \int_0^t {}^\circ Eh(\eta(s))ds$$

for all standard t, as follows. The Riemann integral on the right may be approximated to within any standard $\varepsilon > 0$ by a sum

$$\sum_{J=1}^{k} {}^\circ Eh\eta\left(\frac{j-1}{k}\right)\cdot\frac{1}{k}$$

for k finite, and sufficiently large. By the standard uniform continuity k may also be chosen so that

$$\left|{}^\circ Eh\left(\bar{\eta}\left(\frac{j-1}{k}\right)\right) - Eh\left(\bar{\eta}\left(\frac{i}{m}\right)\right)\right| < \varepsilon$$

whenever $\frac{j-1}{k} \le \frac{i}{m} < \frac{j}{k}$. Thus

$$\left| Eh\left(\bar{\eta}\left(\frac{j-1}{k}\right)\right)\cdot\frac{1}{k} - \sum_{\frac{j-1}{k} < \frac{i}{m} \le \frac{j}{k}} Eh\left(\bar{\eta}\left(\frac{i-1}{m}\right)\right)\cdot\frac{1}{m}\right| < \varepsilon/k$$

and so the difference of the two sides in (18) is no more than 2ε and (18) follows.

We have in fact verified

$$(19) \qquad E\left(\int_0^t h(\bar{\eta}(s))(ds)^{r/q}\right) \approx \begin{cases} 0 & \text{if} \quad r/q = 2,3,\cdots \\ \int_0^t {}^\circ Eh(\bar{\eta}(s))(ds)^1 & r = q \end{cases}$$

and

$$(20) \qquad {}^\circ Eh(\bar{\eta}(t)) = h(\bar{\eta}(0)) + \int_0^t {}^\circ ED_1 h(\bar{\eta}(s))ds$$

From the fact that $\partial_\mu : Z_a \to Z_a$ is a contraction we conclude that there is a constant $M > 1$ depending on a,r and on $\bar{\eta}$ only through $x = \bar{\eta}(0)$ and the coefficients a_{ij} such that for all polynomials h,

$$(21) \qquad |h_{r,\bar{\eta}}|_a \le M|h|_a .$$

We shall now prove that for all $b > M$

$$(22) \qquad |{}^\circ Eh(\bar{\eta}(\cdot))|_b \le |h|_a$$

providing $\bar{\eta}(0) = 0$. The proof will be by induction on the degree of h. If h has degree zero then $°Eh(\bar{\eta}(0)) = h(0)$ and (22) is true. Now assume (22) holds for all polynomials of degree less than that of h. Then by (20)

$$|°Eh(\bar{\eta}(\cdot))|_b = \max\{|h(0)|, |\int_0^1 °ED_1 h(\bar{\eta}(s))ds|_b\} .$$

If $|°Eh(\bar{\eta}(\cdot))|_b = |h(0)|$ then (22) holds. Otherwise

$$|°Eh(\bar{\eta}(\cdot))|_b = |\int_0^1 °ED_1 h(\bar{\eta}(s))ds|_b \leq \frac{1}{b} |°ED_1 h(\eta(\cdot))|_b$$

and by the induction hypothesis we then have $|°Eh(\bar{\eta}(\cdot))|_b < \frac{1}{b} |h_{1,\bar{\eta}}|_a$. Since $b > M$, (22) now follows by (21).

Define translation on $Z_a(n)$ by

$$(T_x f)(y) = f(y+x)$$

for all x,y, in $\mathbb{R}^n$. Then for each x $T_x : Z_a(n) \to Z_a(n)$ is a bounded linear map with operator norm no larger than $e^{na||x||}$.

Define $\bar{\eta}_0 = \bar{\eta} - \bar{\eta}(0)$ so that $\bar{\eta}_0(0) = 0$. For all polynomials h, $h(\bar{\eta}) = (T_x h)(\bar{\eta}_0)$. Consequently,

$$|°Eh(\bar{\eta}(\cdot))|_b \leq e^{a\cdot n\cdot ||\bar{\eta}(0)||} |h|_a \tag{23}$$

for all polynomials h in n variables, and $b > M$, where M depends only on a,q and $\bar{\eta}_0$.

We may now extend E to $h(\bar{\eta}(t))$ for all h in Z_a by polynomial approximation and the inequality (23). Estimate (23) remains valid for this extension.

III. Differential Evolution Equations.

In this section we present a representation for solutions to

$$u_t = Qu \qquad u(x,0) = f(x)$$

where $Q = \sum_{i=1}^m b_i \partial^i$ and $\partial^i = \frac{\partial}{\partial x}$. Here each b_i is a complex constant. Let V be the set of all finite sums of the form

$$x + \sum_r a_r t_r$$

for x in $\mathbb{R}$, a_r in $\mathbb{C}$ and r rational in (0,1]. The flow of Q is a function η from $\mathbb{R}$ in V satisfying

$$\eta(x)(0) = x \text{ ; and} \tag{24}$$

(25) $$(D_1h)(\eta(x)) = Q(h(\eta(x))$$

for all polynomials h and h in $\mathbb{R}$. In [Berger and Sloan, 2] we showed that every such Q has flow. For example,

(i) $\frac{1}{2} a^2\partial^2 + b\partial$ has the flow $x + at_{1/2} + bt_1$;

(ii) $a\partial^3 + b\partial^2 + c\partial$ has the flow $x + (6a)^{1/3} t_{1/3} + \frac{(6a)^{1/3}}{b} t_{2/3} + ct_1$; and

(iii) $\frac{1}{7} a^7\partial^7 + \frac{1}{5} a^5b\partial^6 + \frac{1}{3} a^3b^2\partial^5 + \frac{1}{3} b^3\partial^4$ has the flow $x + at_{1/7} + bt_{2/7}$.

Theorem: If Q has the flow η then for all f in Z(1)

(26) $$u(x,t) = {}^{\circ}Ef(\eta(x)(t))$$

is the solution of $u_t = Qu$, $u(x,0) = f(x)$.

Proof: It is known that a unique solution $u \in Z(2)$ exists for this Cauchy problem.

Fix $s > 0$ and let $\omega(t) = s - t_1(t)$.

For all polynomials h(x,t) it follows directly from the definition of a flow that

$$D_1h(\eta(x)(t),\omega(t)) = (Qh-h_t)(\eta(x)(t,\omega(t)) .$$

Approximating u by polynomials in Z then leads to

$$D_1u(\eta(x)(t),u(t)) = 0 .$$

By (20) we conclude

$${}^{\circ}Eu(\eta(x)(t),\omega(t)) = u(\eta(x(0),\omega(0)) = u(x,s)$$

independent of t.

For all polynomials h(x,t),

$${}^{\circ}Eh(\eta(x)(t),\omega(s)) = {}^{\circ}Eh(\eta(x)(t),0)$$

so polynomial approximation implies

$${}^{\circ}Eu(\eta(x)(s),\omega(s)) = {}^{\circ}Ef(\eta(x)(t)) .$$

Q.E.D.

References.

1. M. Berger and A. Sloan, Radical Differential Calculus, Volume 1, Georgia Tech preprint, December 1979.

2. M. Berger and A. Sloan, A Method of Generalized Characteristics, to appear in Memoirs of the American Mathematical Society.

3. K. Hochberg, A Signed Measure on Path Space Related to Wiener Measure, Annals of Probability, 6, 1978.

HYPERFINITE SPIN MODELS

L. L. Helms
University of Illinois
Urbana, Illinois 61801

1. Introduction. A randomly evolving spin model can be described as follows. At each site of a finite or countable set Γ there is a particle which can be in one of several states making up a set E. The association of states with sites gives rise to a configuration $\eta:\Gamma \to E$ whereby $\eta(x)$, $x \in \Gamma$, is interpreted as the state or spin of the particle at $x \in \Gamma$. Each particle flips randomly among the states in E at random times but not necessarily independently of particles at other sites. Since most physical systems involve only a finite number of particles, the emphasis is on "large but finite" systems. The standard approach makes use of an infinite set of sites as an approximation to a "large but finite" system. We will consider an alternative approach based on a hyperfinite number of sites to describe the probabilistic dynamics of "large but finite" systems.

2. Finite Spin Models. A classical procedure for constructing dynamic probability models is embodied in "Q-matrix" theory (c.f. [2]). Let Q be a real $n \times n$ matrix with entries q_{ij}, $1 \le i, j \le n$, satisfying $q_{ij} \ge 0$ whenever $i \ne j$ and $\sum_j q_{ij} = 0$ for each i. Then for each $t \ge 0$, a stochastic matrix $P(t) = \{p_{ij}(t)\}$ can be defined by putting $P(t) = \exp(tQ)$. The family $\{p_{ij}(t); t \ge 0\}$ can then be used to construct a continuous parameter Markov chain $\{\xi_t; t \ge 0\}$ with state space $\{1,2,\ldots,n\}$. The chain so constructed has the following interpretation in terms of the q_{ij}. Suppose $i \ne j$. If at some given time the chain is in the state i, then the probability that there will be a single jump within a time interval of length Δt to j is $q_{ij}\Delta t + \mathcal{O}(\Delta t)$; the probability of two or more jumps in the interval of length Δt is $\mathcal{O}(\Delta t)$. Since the integers $1,2,\ldots,n$ can be thought of as labels, this construction can be applied to any finite set of states.

Let $\Lambda = Z^d$, the d-dimensional integer lattice, and let Γ be any nonempty finite subset of Λ. The points of Γ will be thought of as particle sites and each particle can be in one of two states 0 and 1. More general sets of states can be used but we will use $\{0,1\}$ to simplify matters. An element $\eta \in S_\Gamma = \{0,1\}^\Gamma$ is called a configuration with $\eta(x)$, $x \in \Gamma$, representing the spin of a particle at $x \in \Gamma$.

In utilizing the above construction, the configurations in S_Γ will play the role of the states of which there are $2^{|\Gamma|}$ in number, where $|\Gamma|$ is the cardinality of Γ. To carry out the construction we need only specify a Q-matrix on $S_\Gamma \times S_\Gamma$. Suppose the spin system is in the configuration $\eta \in S_\Gamma$ at some given time. We will permit only jumps from η to configurations which differ from η at just one site in small time intervals. For each $x \in \Gamma$, let η_x be the configuration obtained from η by changing the spin $\eta(x)$ at x to $1 - \eta(x)$. Transitions of the type $\eta \to \eta_x$, $x \in \Gamma$, will be permitted but transitions $\eta \to \xi$, where ξ differs from η at two or more sites, will not be permitted in small time intervals. In defining a Q-matrix Ω_Γ on $S_\Gamma \times S_\Gamma$, we thus put $\Omega_\Gamma(\eta,\xi) = 0$ whenever ξ differs from η at two or more sites. In order to specify $\Omega_\Gamma(\eta,\eta_x)$ for $x \in \Gamma$ we will let $c(x,\eta)$ be a nonnegative function on $\Gamma \times S_\Gamma$ and put $\Omega_\Gamma(\eta,\eta_x) = c(x,\eta)$ for $x \in \Gamma$. Since the row sums of a Q-matrix must be zero, we complete the definition of Ω_Γ by putting $\Omega_\Gamma(\eta,\eta) = -\sum_{x\in\Gamma} c(x,\eta)$ for each $\eta \in S_\Gamma$. Since $c(x,\eta)$ can depend upon spins at sites other than x, the $c(x,\eta)$ can produce an interaction between spins at different sites. Having defined the Q-matrix Ω_Γ on $S_\Gamma \times S_\Gamma$, we now let $P_\Gamma(t) = \exp(t\Omega_\Gamma)$ for $t \geq 0$ and obtain a family $\{P_\Gamma(t,\eta,\xi);\ t \geq 0\}$ of transition functions on $S_\Gamma \times S_\Gamma$ which in turn define a continuous parameter Markov chain $\{\xi_t;\ t \geq 0\}$ with state space S_Γ and stationary transition function $P_\Gamma(t,\cdot,\cdot)$ on $S_\Gamma \times S_\Gamma$.

Rather than thinking of Ω_Γ as a matrix, we prefer to regard it as the kernel of an operator, denoted by the same symbol, acting on real-valued functions f on S_Γ with $\Omega_\Gamma f(\eta) = \sum_\xi \Omega_\Gamma(\eta,\xi) f(\xi)$. From this point of view

$$(1) \qquad \Omega_\Gamma f(\eta) = \sum_{x\in\Gamma} c(x,\eta)\Delta_x f(\eta), \qquad \eta \in S_\Gamma,$$

where each Δ_x, $x \in \Gamma$, is an operator defined by $\Delta_x f(\eta) = f(\eta_x) - f(\eta)$, $\eta \in S_\Gamma$.

If $c(x,\cdot) \equiv 1$ for all $x \in \Gamma$, the Markov chain ξ_t represents the case of no interaction between particles. In this case, $\Omega_\Gamma f = \sum_{x\in\Gamma}\Delta_x f$ and it follows that $\exp(t\Omega_\Gamma) = \prod_{x\in\Gamma}\exp(t\Delta_x)$ since the operators Δ_x are easily seen to commute. Suppose now that for each $x \in \Gamma$, f_x is a function on S_Γ which depends only upon $\eta(x)$. Then $P_\Gamma(t)(\prod_{x\in\Gamma} f_x) = \prod_{x\in\Gamma} P_{\{x\}}(t) f_x$ where $P_{\{x\}}(t) = \exp(t\Delta_x)$. This implies that the transition density $P_\Gamma(t,\eta,\xi)$ is the product of the densities $P_{\{x\}}(t,\eta(x),\xi(x))$, $x \in \Gamma$. By examining the construction of the probability measures P_Γ^η, $\eta \in S_\Gamma$, from the transition densities, it is easily seen that the processes $\{\xi_{\cdot}(x); x \in \Gamma\}$ are independent stochastic processes. This justifies the statement that $c(x,\cdot) \equiv 1$ case corresponds to no interaction between sites, not just at a fixed time but at different times for different sites.

3. Infinite Spin Systems. The standard model of a "large but finite" spin system entails placing a particle at each site of $\Lambda = Z^d$. Again $\eta \in S = \{0,1\}^{\Lambda}$ are called configurations. If $\{x_n\}$ is an enumeration of the elements of Λ and we define $\rho(\eta,\eta') = \sum_{n=1}^{\infty} 2^{-n}|\eta(x_n) - \eta'(x_n)|$ for $\eta,\eta' \in S$, then (S,ρ) is a compact metric space. As usual, $C(S)$ will denote the Banach space of real continuous functions on S with the supremum norm. The set T of tame functions consists of those functions f on S for which $f(\eta)$ depends only upon $\eta|_J$, the restriction of η to J, for some finite $J \subset \Lambda$. Suppose now that $\{c(x,\cdot); x \in \Lambda\}$ is a subset of $C(S)$ such that

(2) $$0 \leq c(x,\cdot) \leq M < +\infty, \qquad x \in \Lambda,$$

for some constant M. By analogy with the finite case, we can define an operator Ω on T by putting

(3) $$\Omega f(\eta) = \sum_{x\in\Lambda} c(x,\eta)\Delta_x f(\eta), \qquad f \in T,\ \eta \in S.$$

Under a suitable condition limiting the influence of remote sites of Λ on each $c(x,\cdot)$, namely

$$\sup_{x\in\Lambda} \sum_{y\in\Lambda} \sup_{\eta\in S}|c(x,\eta_y) - c(x,\eta)| < +\infty,$$

T. Liggett [5] has shown, using the Hille-Yosida Theorem, that the generally unbounded operator Ω has a closed extension which is the generator of a unique Feller semigroup of operators on $C(S)$. Generally speaking, a satisfactory standard global description of an infinite spin system is lacking in the case of nonnegative, uniformly bounded, and continuous interaction functions.

Consider once again the special case $c(x,\cdot) \equiv 1$ for all $x \in \Lambda$ and let $\{P(t); t \geq 0\}$ be the unique Feller semigroup on $C(S)$ generated by the corresponding Ω. Let Γ be any nonempty finite subset of Λ and let $T(\Gamma)$ be the set of cylindrical functions on S with base Γ. By considering the resolvent equations for Ω and the Ω_Γ of the preceding section, it is easily seen that $P(t)f = P_\Gamma(t)f$ for $t \geq 0$, $f \in T(\Gamma)$, where $P_\Gamma(t)$ is the semigroup corresponding to

$c(x,\cdot) \equiv 1$, $x \in \Gamma$. This implies that the probability measures P^η, $\eta \in S$, constructed on an appropriate sample space (defined below) using the $P(t)$ semigroup and initial configuration $\eta \in S$ are product measures $\prod_{x\in\Lambda} P^{\eta(x)}_{\{x\}}$. The Markov process $\{\xi_t; t \geq 0\}$ governed by P^η has the property that the processes $\{\xi_\cdot(x);\ x \in \Lambda\}$ are independent processes.

The appropriate sample space for infinite spin systems is defined as follows. The sample space is the set $D = D[0,\infty)$ of all maps $\omega: [0,\infty) \to S$ which have left limits and are right-continuous. The coordinate functions ξ_t, $t \geq 0$, on D are defined by $\xi_t(\omega) = \omega(t)$ and the appropriate σ-fields are $F_t = \sigma(\xi_s;\ s \leq t)$ and $F = \sigma\{\xi_s;\ s \geq 0\}$. We will need the Skorohod J_1-topology on $D = D[0,\infty)$ which is defined as follows. A sequence $\{\omega_n\}$ in D converges to $\omega \in D$ in the J_1-topology if there is a sequence of continuous one-to-one mappings $\{\lambda_n(t)\}$ of the interval $[0,\infty)$ onto itself such that for each positive integer $m \in N$

$$\lim_{n\to\infty}\Big[\sup_{0\leq t\leq m} \rho(\omega_n(t),\omega(\lambda_n(t))) + \sup_{0\leq t\leq m} |\lambda_n(t) - t|\Big] = 0$$

Since most of the literature pertaining to the Skorohod topology does not apply to paths on unbounded intervals but rather to paths on bounded intervals, we will make use of the following device which is due to C. J. Stone [9]. A new state space $\tilde{S}$ is constructed by adjoining an ideal element, corresponding to $t = 1$, to the space-time set $S \times [0,1)$ and a metric $\tilde{\rho}$ on $\tilde{S}$ is defined so that $(\tilde{S},\tilde{\rho})$ is a compact metric space. The space $\tilde{D}[0,1]$ is then defined as before and has the property that any path in this new space is continuous at $t = 1$. A one-to-one and onto transformation T from $D[0,\infty)$ to $\tilde{D}[0,1]$ is defined so that a sequence $\{\omega_n\}$ in $D[0,\infty)$ converges to an element $\omega \in D[0,\infty)$ in the J_1-topology if and only if the sequence $\{T\omega_n\}$ in $\tilde{D}[0,1]$ converges in the J_1-topology to $T\omega \in \tilde{D}[0,1]$. The Skorohod J_1-topology on $D[0,\infty)$ is clearly equivalent to a metric topology, given by a metric s, in which $D[0,\infty)$ is a complete separable metric space. Moreover, a necessary and sufficient condition is known for determining if a subset of $\tilde{D}[0,1]$ is conditionally compact in the J_1-topology (c.f. [8]; the results therein are stated for paths taking on values in R but are valid for paths taking on values in a compact metric space). The compactness criterion for $\tilde{D}[0,1]$ can be translated to $D[0,\infty)$ as follows For each $\omega \in D[0,\infty)$ and $\delta > 0$ define the total oscillation of ω on $[0,1/\delta]$ by

$$0_\omega(\delta) = \max[\sup_{\substack{t-\delta\le t'\le t\le t''\le t+\delta \\ 0\le t\le 1/\delta}} (\min(\rho(\omega(t'),\omega(t)),\rho(\omega(t''),\omega(t)))),$$

$$\sup_{0\le t\le\delta} \rho(\omega(t),\omega(0))].$$

Then $K \subset D[0,\infty]$ is conditionally compact in the J_1-topology if and only if

(4) $$\lim_{\delta\to 0} \sup_{\omega\in K} 0_\omega(\delta) = 0.$$

Letting B denote the class of Baire subsets of $D[0,\infty)$, the smallest σ-field relative to which all the bounded real continuous functions on $D[0,\infty)$ are measurable, we can also utilize $\tilde{D}[0,1]$ to show that $B = F = \sigma(\xi_s ; s \ge 0)$. For each $t \in [0,1]$ define $\tilde{\xi}_t$ to be the appropriate coordinate function on $\tilde{D}[0,1]$ and let $\tilde{F} = \sigma(\tilde{\xi}_t ; 0 \le t \le 1) = \sigma(\tilde{\xi}_t ; 0 \le t < 1)$ (the coordinate function $\tilde{\xi}_1$ is constant). Letting $\tilde{B}$ denote the class of Baire subsets of $\tilde{D}[0,1]$, we have $\tilde{B} = \tilde{F}$. The proof of the fact that $\tilde{B} \subset \tilde{F}$ is the same as that given in [8]. The proof that $\tilde{F} \subset \tilde{B}$ in [8], however, requires a minor change. Fix $t \ge 0$ and note that $\lim_{h\downarrow 0}(1/h)\int_t^{t+h} f(\tilde{\xi}_s)ds = f(\tilde{\xi}_t)$ for each $f \in C(\tilde{S})$ by the right continuity of paths. Since each of the integrals $\int_t^{t+h} f(\tilde{\xi}_s)ds$ is known to be continuous in the J_1-topology whenever $f \in C(\tilde{S})$, $f(\tilde{\xi}_t)$ is a limit of J_1-continuous functions on $\tilde{D}[0,1]$. This means that each $\tilde{\xi}_t$, $0 \le t \le 1$, is measurable relative to $\tilde{B}$ and $\tilde{F} \subset \tilde{B}$. By examining the transformation T from $D[0,\infty)$ onto $\tilde{D}[0,1]$ used in [9], it is easily seen that $B = F$.

4. <u>Hyperfinite Spin Models</u>. Consider now a denumerably comprehensive enlargement of a structure containing a countable base for the J_1-topology on D and the real numbers R (see [11] for terminology and notation). If $\gamma \in {}^*N \sim N$ is a fixed infinite integer, let $\Gamma = \{x \in {}^*\Lambda ; x = (x_1,\ldots,x_d) \in {}^*Z^d$ and $\max_{1\le i\le d} |x_i| \le \gamma\}$ and let S_Γ be the set of internal maps from Γ into $\{0,1\}$. If η is an internal function with domain containing Γ and taking on values in $\{0,1\}$ and $\xi \in {}^*S$, $[\eta,\xi]_\Gamma$ will denote the element of ${}^*S$ which is equal to η on Γ and ξ on ${}^*\Lambda \sim \Gamma$. Let $\{c(x,\cdot) ; x \in \Lambda\}$ be a subset of $C(S)$ satisfying (2) and let φ be a fixed element of ${}^*S$ which will serve as a boundary condition outside Γ. Letting ${}^*c$ denote the non-standard extension of the function c, as a function on $\Lambda \times S$, we can define an internal operator $\Omega_{\Gamma,\varphi}$ on ${}^*C(S)$ analogous to the operator Ω

given by (1) by putting

$$(5) \qquad \Omega_{\Gamma,\varphi} f(\eta) = \sum_{x\in\Gamma} {}^*c(x,[\eta,\varphi]_\Gamma)\,\Delta_x f(\eta)$$

for $\eta \in {}^*S$ and $f \in {}^*C(S)$. Clearly, $\Omega_{\Gamma,\varphi}$ is an internally bounded operator on ${}^*C(S)$ with $\|\Omega_{\Gamma,\varphi}\| \leq 2M|\Gamma|$, where $|\Gamma|$ is the internal cardinality of Γ, and $\Omega_{\Gamma,\varphi} 1 = 0$. We can therefore define an internal Markov semigroup of operators $\{T_{\Gamma,\varphi}(t) \; ; \; t \in {}^*[0,\infty)\}$ on ${}^*C(S)$ by putting $T_{\Gamma,\varphi}(t) = \exp(t\Omega_{\Gamma,\varphi})$. The generator $\Omega_{\Gamma,\varphi}$ defined by (5) has the interpretation that the spin at $x \in \Gamma$ will flip in a time interval of length Δt with probability ${}^*c(x,[\eta,\varphi]_\Gamma)\Delta t + \mathcal{O}(\Delta t)$ whereas the spin at a site $x \notin \Gamma$ will flip with probability $\mathcal{O}(\Delta t)$. We know, using the transfer principle, that to each $\sigma \in {}^*S$ there corresponds an internal probability measure $\overline{P}_c^\sigma$ on ${}^*D$ such that $\{{}^*D, {}^*F, {}^*F_t, {}^*\xi_t, \overline{P}_c^\sigma\}_{\sigma\in {}^*S, t\in {}^*[0,\infty)}$ is an internal Markov process. Note that ${}^*\xi_t$ is the nonstandard extension of $\xi_t(\omega)$ as a function on $D \times [0,\infty)$. Denoting expected values relative to $\overline{P}_c^\sigma$ by $\overline{E}_c^\sigma(\cdot)$, $\overline{E}_c^\sigma[f({}^*\xi_t)] = T_{\Gamma,\varphi}(t)f(\sigma)$ whenever $\sigma \in {}^*S$, $t \in {}^*[0,\infty)$, and $f \in {}^*C(S)$. We also know from the transfer principle that ${}^*\xi_t$ remains fixed and equal to φ on ${}^*\Lambda \sim \Gamma$ with $\overline{P}_c^\sigma$ probability one for each $\sigma \in {}^*S$ provided σ is equal to φ on ${}^*\Lambda \sim \Gamma$. For this reason, we will consider initial states $[\eta,\varphi]_\Gamma$ for $\eta \in S_\Gamma$ with associated probabilities denoted by $\overline{P}_c^{[\eta,\varphi]}$. These internal probabilities describe the time evolution of spins at sites $x \in \Gamma$ with φ serving as a fixed boundary condition on ${}^*\Lambda \sim \Gamma$. The state space of the ${}^*\xi_t$ process is essentially the internal set S_Γ whenever the initial state is $[\eta,\varphi]_\Gamma$ for some $\eta \in S_\Gamma$. We will next show how the measures $\overline{P}_c^{[\eta,\varphi]}$ can be used to induce standard measures on D.

Consider the space $D = D[0,\infty)$ and the topology T_D defined by the Skorohod metric. If $\omega \in D$, the monad of ω is defined to be $m(\omega) = \bigcap_{\omega\in O\in T_D} {}^*O$. The standard part relation on ${}^*D$ is defined by $\omega = \text{st}\,\omega'$ if $\omega' \in m(\omega)$. The near standard points of ${}^*D$, $ns({}^*D)$, are those $\omega' \in {}^*D$ which belong to $m(\omega)$ for some $\omega \in D$. Now let ν be an internal probability measure on $({}^*D, {}^*B)$, where ${}^*B$ is the set of internal Baire subsets of ${}^*D$, and let ν_o be the Loeb measure on the external σ-algebra $\sigma({}^*B)$. R. M. Anderson and S. Rashid [1] (see also [6, 7]) have shown that if $\nu_o({}^*D \sim ns({}^*D)) = 0$, then there is a probability measure ν_D on (D,B) such that

$$(6) \qquad \int f\, d\nu_D = \int {}^\circ({}^*f)\,d\nu_o \simeq \int {}^*f\, d\nu$$

whenever $f \in C(D)$ ("$a \simeq b$" means $a - b \in m(0)$). An internal measure ν satisfying the hypothesis of this statement is said to be near-standardly concentrated on $^*D$. To prove that an internal probability measure ν is near-standardly concentrated it suffices to show that for each standard $\varepsilon > 0$ there is a compact set $K \subset D$ such that $\nu(^*K) \geq 1 - \varepsilon$. In order to apply this condition to the internal measures $\overline{P}_c^{[\eta,\varphi]}$, we will first establish a reference family of near-standardly concentrated measures, namely the $\overline{P}_1^{[\eta,\varphi]}$ for which $c \equiv 1$.

Lemma 1. For each $\eta \in S$, $\overline{P}_1^{[^*\eta,\varphi]}$ is near-standardly concentrated on $^*D$.

Proof: Fix $\eta \in S$. The internal measure $\overline{P}_1^{[^*\eta,\varphi]}$ corresponds to the internal Markov semigroup $\{T_{\Gamma,\varphi}(t);\ t \in {}^*[0,\infty)\}$ with internal generator $\Omega_{\Gamma,\varphi} f(\xi) = \sum_{x \in \Gamma} \Delta_x f(\xi)$, $\xi \in {}^*S$, $f \in {}^*C(S)$, which is independent of φ and corresponds to an internal spin system for which the spins at sites $x \in \Gamma$ act independently of each other and the spins at sites $x \notin \Gamma$ remain at $\varphi(x)$. Note that the $T_{\Gamma,\varphi}(t)$ semigroup and the measure $\overline{P}_1^{[^*\eta,\varphi]}$ are uniquely determined by the generator $\Omega_{\Gamma,\varphi}$ and the initial configuration $[^*\eta,\varphi]$. On the other hand, a probability measure P_1^η on D can be constructed directly using Liggett's result, having generator $\Omega f(\xi) = \sum_{x \in \Lambda} \Delta_x f(\xi)$, $\xi \in S$, $f \in T$; in fact, P_1^η is the product measure $\prod_{x \in \Lambda} P_{\{x\}}^{\eta(x)}$ where each $P_{\{x\}}^{\eta(x)}$ is a probability measure determining the time evolution of the spin at site $x \in \Lambda$. Since D is a complete separable metric space, given $\varepsilon > 0$ in R there is a compact set $K \subset D$ such that $P_1^\eta(K) > 1 - \varepsilon$ (c.f. [8]). Consider $^*P_1^\eta$, the nonstandard extension of P_1^η, which is the product measure $\prod_{x \in {}^*\Lambda} {}^*P_{\{x\}}^{\eta(x)}$. For each $x \in {}^*\Lambda \sim \Gamma$, let μ_x be an internal probability measure governing the time evolution of a spin at x which remains at $\varphi(x)$. Clearly, $\overline{P}_1^{[^*\eta,\varphi]} = (\prod_{x \in \Gamma} {}^*P_{\{x\}}^{\eta(x)}) \times (\prod_{x \in {}^*\Lambda \sim \Gamma} \mu_x)$. Consider now the internal set

$$({}^*K)_\Gamma = \{\omega \in {}^*D : \exists \omega' \in {}^*K \ni \omega(t)|_\Gamma = \omega'(t)|_\Gamma\ \forall t \in {}^*[0,\infty)\}$$

which belongs to $^*F$. Then $\overline{P}_1^{[^*\eta,\varphi]}(({}^*K)_\Gamma) \geq {}^*P_1^\eta({}^*K) > 1 - \varepsilon$. We will now show that $({}^*K)_\Gamma \subset ns({}^*D)$. If $\omega \in ({}^*K)_\Gamma$, then there is an $\omega' \subset {}^*K$ such that $\omega(t)|_\Gamma = \omega'(t)|_\Gamma$ for all $t \in {}^*[0,\infty)$ and therefore ${}^*s(\omega,\omega') \simeq 0$. Since K is compact and $\omega' \in {}^*K$, there is an $\omega_o \in K$ such that ${}^*s(\omega',\omega_o) \simeq 0$; thus, ${}^*s(\omega,\omega_o) \simeq 0$ and $\omega \in ns({}^*D)$. This proves that the Loeb measure of ${}^*D \sim ns({}^*D)$ is zero.

To get at the case of general c, we must look at an operation on compact subsets of D which can be described as follows. Let L_M be the set of mappings $\alpha:[0,\infty) \times D \to [0,\infty)^{\Lambda}$ (that is, vector-valued functions $\alpha = \langle \alpha_x(t,\omega); x \in \Lambda \rangle$ for which each $\alpha_x(\cdot,\omega)$ is nondecreasing on $[0,\infty)$, $\alpha_x(0,\omega) = 0$, and $|\alpha_x(t,\omega) - \alpha_x(t',\omega)| \leq M|t - t'|$ whenever $\omega \in D$, $0 \leq t, t' < +\infty$. If $\omega \in D$ and $\alpha \in L_M$, $\omega \circ \alpha$ will denote the element of D defined by $(\omega \circ \alpha)(t) = \langle \omega_x(\alpha_x(t,\omega)); x \in \Lambda \rangle$. If $A \subset D$, we let $A_M = \{\omega \circ \alpha : \omega \in A$ and $\alpha \in L_M\}$. Note that $\omega \circ \alpha_x$ denotes the element of D for which all components of ω are composed with $\alpha_x(t)$.

<u>Lemma 2</u>. If K is a conditionally compact subset of D, then K_M is a conditionally compact subset of D.

Proof: We need only verify that K_M satisfies (4). If $\omega' \in K_M$, then $\omega' = \omega \circ \alpha$ for some $\omega \in K$, $\alpha \in L_M$. Fix $\varepsilon > 0$ in R and choose $m \in N$ such that $\sum_{n=m+1}^{\infty} 2^{-n} < \varepsilon$. Then for any $t, t' \in [0,\infty)$,

$$\rho(\omega \circ \alpha(t), \omega \circ \alpha(t')) \leq \sum_{n=1}^{m} 2^{-n} |\omega_{x_n}(\alpha_{x_n}(t)) - \omega_{x_n}(\alpha_{x_n}(t'))| + \varepsilon$$

$$\leq \sum_{n=1}^{m} \rho(\omega \circ \alpha_{x_n}(t), \omega \circ \alpha_{x_n}(t')) + \varepsilon$$

Clearly, $0_{\omega'}(\delta) = 0_{\omega\circ\alpha}(\delta) \leq m\, 0_{\omega}(M\delta) + \varepsilon$ and (4) is satisfied.

The proof of the following theorem is based on the concept of simultaneous random changes of time scales of the component processes ${}^*\xi_{\cdot}(x)$, $x \in \Gamma$. A random change of time scale for a single process was first introduced by Volkonskii [12] and simultaneous random changes of time scales by the author in [3] for vector Markov processes with finitely many components.

We can now treat the general case of speed functions $c(x,\cdot)$ satisfying (2).

<u>Theorem 3</u>. For each $\eta \in S$, $\overline{P}_c^{[{}^*\eta,\varphi]}$ is near-standardly concentrated on ${}^*D$.

Proof: Using the results of [3] via the transfer principle, there is a $\tau \in {}^*L_M$ such that $\tau_x \equiv 0$ for $x \in {}^*\Lambda \sim \Gamma$ and

$$\tau_x(t,\omega) = \int_0^t {}^*c(x,[{}^*\xi_{\tau(s)}(\omega),\varphi]_{\Gamma})\,{}^*ds, \quad t \in {}^*[0,\infty).$$

where ${}^*\xi_{\tau(s)}(\omega) = \langle \omega_x(\tau_x(s,\omega));\ x \in {}^*\Lambda \rangle$, $\omega \in {}^*D$; moreover, the interna

Markov process $\eta_t = {}^*\xi_{\tau(t)}$ has the property that $\bar{E}_1^{[\eta,\varphi]}[f(\eta_t)], f \in {}^*C(S)$, defines an internal Feller semigroup having internal generator $\Omega_{\Gamma,\varphi}$. This means that

$$\bar{E}_c^{[\eta,\varphi]}[f({}^*\xi_t)] = \bar{E}_1^{[\eta,\varphi]}[f(\eta_t)] = \bar{E}_1^{[\eta,\varphi]}[f({}^*\xi_{\tau(t)})]$$

for $t \in {}^*[0,\infty)$, $f \in {}^*C(S)$, and $\eta \in {}^*S$. Using the Markov properties of the ${}^*\xi_t$ and ${}^*\xi_{\tau(t)}$ processes internally

$$\bar{E}_c^{[\eta,\varphi]}[\prod_{i=1}^n f_i({}^*\xi_{t_i})] = \bar{E}_1^{[\eta,\varphi]}[\prod_{i=1}^n f_i({}^*\xi_{\tau(t_i)})]$$

whenever $\{t_i\}$ is a hyperfinite subset of ${}^*[0,\infty)$ and $f_i \in {}^*C(S)$, $i = 1,\dots,n$. Consider the internal mapping ${}^*\xi_\tau$ which takes $\omega \in {}^*D$ into ${}^*\xi_{\tau(\cdot)}(\omega) \in {}^*D$. The last equation implies that $\bar{P}_c^{[\eta,\varphi]} = \bar{P}_1^{[\eta,\varphi]} \circ {}^*\xi_\tau^{-1}$, $\eta \in {}^*S$. By Lemma 1, given $\varepsilon > 0$ in R there is a compact set $K \subset D$ such that $\bar{P}_1^{[{}^*\eta,\varphi]}({}^*K) \geq 1 - \varepsilon$. By Lemma 2, the closure $\bar{K}_M$ of K_M in D is compact in D. Since ${}^*\xi_\tau {}^*K \subset {}^*K_M \subset {}^*\bar{K}_M$, ${}^*K \subset {}^*\xi_\tau^{-1} {}^*\xi_\tau {}^*K \subset {}^*\xi_\tau^{-1} {}^*\bar{K}_M$ and

$$\bar{P}_c^{[{}^*\eta,\varphi]}({}^*\bar{K}_M) = \bar{P}_1^{[{}^*\eta,\varphi]}({}^*\xi_\tau^{-1} {}^*\bar{K}_M) \geq \bar{P}_1^{[{}^*\eta,\varphi]}({}^*K) \geq 1 - \varepsilon.$$

This proves that $\bar{P}_c^{[{}^*\eta,\varphi]}$ is near-standardly concentrated on ${}^*D$.

For each $\eta \in S$, let $\bar{P}_{c,o}^{[{}^*\eta,\varphi]}$ denote the Loeb measure determined by $\bar{P}_c^{[{}^*\eta,\varphi]}$ on the external σ-field $\sigma({}^*F)$. Since each of the internal probability measures $\bar{P}_c^{[{}^*\eta,\varphi]}$ is near-standardly concentrated on ${}^*D$, each induces a probability measure $P_c^{\eta,\varphi}$ on (D,B) for which (6) holds. Consider the system $\{D,F,F_t,\xi_t,P_c^{\eta,\varphi}\}_{\eta\in S,t\in[0,\infty)}$, which we might take as a model for an infinite spin system with interaction function $c(x,\cdot)$ on $\Lambda \times S$. Although it can not be shown at this time that the ξ_t process is a Markov process relative to each $P_c^{\eta,\varphi}$, there are many analytical properties of the internal semigroup $T_{\Gamma,\varphi}(t)$ which can be standardized. We will consider one such property to illustrate the point.

Consider a fixed $f \in {}^*C(S)$ and $\bar{P}_c^{[\eta,\varphi]}$, $\eta \in S_\Gamma$. It is known that

$$f({}^*\xi_t) - \int_0^t \Omega_{\Gamma,\varphi} f({}^*\xi_s)\, {}^*ds$$

is an internal ${}^*F_t$-martingale relative to $\bar{P}_c^{[\eta,\varphi]}$, where ${}^*ds$ denotes internal Lebesgue measure on ${}^*R$. This fact can be used to show that the $\{P_c^{\eta,\varphi}; \eta \in S\}$ solve the martingale problem for the operator Ω (c.f. [10]); that is, for each $\eta \in S$ and $f \in T$, $P_c^{\eta,\varphi}[\xi_o = \eta] = 1$ and

$$(7) \qquad Y_t^f = f(\xi_t) - \int_0^t \Omega f(\xi_s)\,ds$$

is an F_t-martingale relative to $P_c^{\eta,\varphi}$.

We can not use (6) when integrating the random variables Y_t^f defined by (7) since the first term $f(\xi_t(\omega))$ in (7) is not continuous relative to the Skorohod topology on D (in fact, the coordinate function ξ_t is not continuous). This lack of continuity can be eliminated by time averaging as follows. For any $f \in C(S)$ for which $\Omega f \in C(S)$, $t \in [0,\infty)$, and $h > 0$, let

$$Y_{t,t+h}^f(\omega) = (1/h)[\int_t^{t+h} f(\xi_u)du - \int_t^{t+h}\int_0^u \Omega f(\xi_v(\omega))dv\,du]$$

Each such $Y_{t,t+h}^f$ is continuous in the Skorohod topology on D and $\lim_{h\downarrow 0} Y_{t,t+h}^f = Y_t^f$ by right-continuity of paths whenever $f \in C(S)$. Another continuity problem has to do with the fact that we must integrate over sets $E \in \sigma(\xi_s; s \leq t)$ in order to verify that the Y_t^f are martingales. To circumvent this problem, for each $t > 0$ let $C_t(D)$ be the set of bounded real continuous functions g on D such that $g(\omega)$ depends only upon $\omega|_{[0,t]}$ and let B_t be the smallest σ-field of subsets of D relative to which each $g \in C_t(D)$ is measurable. It is easil seen that each coordinate function ξ_t is measurable relative to $B_{t+} = \bigcap_{s>t} B_s$.

Lemma 4. The family of probability measures $\{P^\eta; \eta \in S\}$ solves the martingale problem for Ω if $P^\eta[\xi_o = \eta] = 1$ for each $\eta \in S$ and

$$(8) \qquad \int g\, Y_{t,t+\Delta t}^f dP^\eta = \int f\, Y_{s,s+\Delta s}^f dP^\eta$$

for all $g \in C_s(D)$, $f \in T$, and $0 \leq s < s + \Delta s < t < t + \Delta t < \infty$.

Proof: Fix $\eta \in S$. By letting Δs, Δt in (8) tend to zero and then replacing g by the indicator function of a set, we have $\int_E Y_t^f dP^\eta = \int_E Y_s^f dP^\eta$ for all $E \in B_s$. Suppose $s < u < t$ and $E \in \sigma(\xi_v; v \leq s)$. Then $E \in B_{s+} \subset B_u$ and $\int_E Y_t^f dP^\eta = \int_E Y_u^f dP^\eta$. Now letting u decrease to s, we see that Y_t^f is an F_t-martingale for P^η.

Corollary 5. The family $\{P_c^{\eta,\varphi};\ \eta \in S\}$ solves the martingale problem for the operator Ω.

Proof: Consider the standard objects $f \in T$, $0 \leq s < s + \Delta s < t < t + \Delta t < \infty$, $\eta \in S$, and $g \in C_s(D)$. Since $\{\overline{P}_c^{[\eta,\varphi]}; \eta \in {}^*S\}$ solves the martingale problem for $\Omega_{T,\varphi}$,

$$(9)\qquad (1/\Delta t)\bar{E}_c^{[^*\eta,\varphi]}[^*g(\int_t^{t+\Delta t}\{^*f(^*\xi_u) - \int_0^u \Omega_{\Gamma,\varphi}{}^*f(^*\xi_v)^*dv\}^*du)]$$

$$= (1/\Delta s)\bar{E}_c^{[^*\eta,\varphi]}[^*g(\int_s^{s+\Delta s}\{^*f(^*\xi_u) - \int_0^u \Omega_{\Gamma,\varphi}{}^*f(^*\xi_v)^*dv\}^*du)].$$

Since (6) implies that

$$(1/\Delta t)\bar{E}_c^{[^*\eta,\varphi]}[^*g(\int_t^{t+\Delta t}\{^*f(^*\xi_u) - \int_0^u {}^*(\Omega f)(^*\xi_v)^*dv\}^*du)]$$

$$(10)\qquad \simeq (1/\Delta t)E_c^{\eta,\varphi}[g(\int_t^{t+\Delta t}\{f(\xi_u) - \int_0^u \Omega f(\xi_v)dv\}du)]$$

it suffices to show, for example, that the values on the left in (9) and (10) are infinitesimally close. Suppose the tame function $f(\eta)$ depends only upon $\eta|_J$ where J is finite subset of Λ. Since for $\eta \in {}^*S$

$$|\Omega_{\Gamma,\varphi}{}^*f(\eta) - {}^*(\Omega f)(\eta)| \le \sum_{x\in J} |{}^*c(x,[\eta,\varphi]_\Gamma) - {}^*c(x,\eta)|\,|\Delta_x{}^*f(\eta)|$$

$$\le 2\|f\| \sum_{x\in J} |{}^*c(x,[\eta,\varphi]_\Gamma) - {}^*c(x,\eta)|$$

and the latter sum is infinitesimal by the continuity of each $c(x,\cdot)$, $x \in J$, equation (8) is satisfied. We need only show that $P_c^{\eta,\varphi}[\xi_o = \eta] = 1$. We know by the transfer principle that $\bar{P}_c^{[^*\eta,\varphi]}[^*\xi_o = [^*\eta,\varphi]] = 1$ and therefore $\bar{P}_{c,o}^{[^*\eta,\varphi]}[(^*\xi_o = [^*\eta,\varphi]) \cap ns(^*D)] = 1$. Consider any ω in $(^*\xi_o = [^*\eta,\varphi]) \cap ns(^*D)$. Then there is an $\omega' \in D$ such that ${}^*s(\omega,\omega') \simeq 0$ and, in particular, ${}^*\rho(\omega(0),\omega'(0)) \simeq 0$. This implies that $\omega(0)$ and $\omega'(0)$ agree on the standard points of Γ so that $\omega \in st^{-1}(\xi_o = \eta)$ and $(^*\xi_o = [^*\eta,\varphi]) \cap ns(^*D) \subset st^{-1}(\xi_o = \eta)$. Thus, $P_c^{\eta,\varphi}[\xi_o = \eta] = P_{c,o}^{[^*\eta,\varphi]}[st^{-1}(\xi_o = \eta)] = 1$.

The above standardization of each internal probability measure $\bar{P}_c^{[^*\eta,\varphi]}$ to produce a standard probability measure $P_c^{\eta,\varphi}$ on (D,F) is a pointwise operation. The assertion that $E_c^{\eta,\varphi}[f(\xi_t)]$, $t > 0$, $f \in C(S)$, is a measurable function of η is global and can not be justified at this time except under stringent conditions on the interaction functions $c(x,\cdot)$ (see [4]). Any speculation about the measurability of $E_c^{\eta,\varphi}[f(\xi_t)] = {}^\circ[E_c^{[^*\eta,\varphi]}[f(^*\xi_t)]$, as a function of η, must be tempered by the well-known fact that ${}^\circ[\sin \omega x]$ is not a measurable function of x for suitably chosen $\omega \in {}^*N \sim N$ (see [11]).

REFERENCES

1. R. M. Anderson and S. Rashid, A nonstandard characterization of weak convergence. Proc. Amer. Math. Soc., 69(1978), 327-332.

2. J. L. Doob, Stochastic Processes, Wiley, New York, 1953.

3. L. Helms, Ergodic properties of several interacting Poisson particles, Adv. in Math., 12(1974), 32-57.

4. L. Helms and P. A. Loeb, Applications of nonstandard analysis to spin models. Jour. of Math. Analysis and Applic., 69(1979), 341-352.

5. T. M. Liggett, Existence theorems for infinite particle systems, Trans. Amer. Math. Soc., 165(1972), 471-481.

6. P. A. Loeb, An Introduction to Nonstandard Analysis and Hyperfinite Probability Theory, in Probabilistic Analysis and Related Topics, A. T. Bharucha-Reid, Ed., Vol. 2, pp. 105-141, Academic Press, 1969.

7. D. W. Müller, Nonstandard proofs of invariance principles in probability theory, in Applications of Model Theory to Algebra, Analysis, and Probability, W. A. J. Luxemburg, ed., pp. 186-194, Holt, Rinehart and Winston, 1969.

8. K. R. Parthasarathy, Probability Measures on Metric Spaces, Academic Press, New York, 1967.

9. C. J. Stone, Weak convergence of stochastic processes defined on semi-infinite time intervals, Proc. Amer. Math. Soc., 14(1963), 694-696.

10. D. W. Stroock, Lectures on Infinite Interacting Systems, Kyoto University, Kinokuniya Book-Store Co., Ltd., Tokyo, 1978.

11. K. D. Stroyan and W. A. J. Luxemburg, Introduction to the Theory of Infinitesimals, Academic Press, New York, 1976.

12. V. A. Volkonskii, Random substitution of time in strong Markov processes, Theory Prob. Appl., 3(1958), 310-325.

NONSTANDARD ANALYSIS AND THE THEORY OF BANACH SPACES

C. Ward Henson
Department of Mathematics
University of Illinois at Urbana/Champaign
Urbana, IL 61801

L. C. Moore, Jr.
Department of Mathematics
Duke University
Durham, NC 27706

Introduction.

Nonstandard analysis has proved to be a natural framework for studying the local properties of Banach spaces. The central construction in this approach is the nonstandard hull, introduced by Luxemburg [LUX 1]. Not only is this a useful tool in studying the local theory of Banach space geometry, but also nonstandard hulls arise naturally in many other places within nonstandard analysis.

In this survey paper we have tried to present two kinds of results: (A) representative examples of important standard theorems about Banach spaces, to illustrate the type of result the nonstandard methods seem best able to handle; (B) detailed discussion and examples of the various tools and ideas that are now available within nonstandard analysis for application within Banach space theory and related areas of functional analysis. Necessarily we have assumed that the reader has some beginning familiarity with nonstandard analysis and with Banach space theory. However, we have attempted to keep these prerequisites to a minimum and we hope that our exposition will be useful both to Banach space theorists and to nonstandard analysts.

Many results discussed here were originally proved using ultraproducts. This approach is essentially equivalent to the use of nonstandard analysis (as is discussed below and in the introduction to [HEN 3].) The survey paper by S. Heinrich [HEI 4] together with this paper should give a reasonable complete account of the uses of these model-theoretic methods in Banach space theory.

The origins of these methods lie in the papers by W. A. J. Luxemburg [LUX 1], where nonstandard hulls were introduced, and by D. Dacunha-Castelle and J.-L. Krivine [DCK 1], where Banach space ultraproducts were first used. (See also [BDCK].) Even earlier

---

Research supported by the National Science Foundation through Grants MCS-8003251 and MCS-8002183.

A. Robinson had used nonstandard analysis in representing the dual space of ℓ_∞ [ROB 1].

Among the standard Banach space theorems which have been first proved using the methods described here are the following: S. Heinrich's proof that the Uniform Approximation Property UAP is self-dual, and other results about the UAP (see Theorem 7.7); C. W. Henson's positive solution of a conjecture due to P. Enflo and H. P. Rosenthal concerning the local geometry of L_p-spaces (see Theorem 7.8); J.-L. Krivine's result that every infinite dimensional sequence in a Banach space has block finitely represented in it a sequence which is isometric to the usual basis of c_0 or of ℓ_p for some p $(1 \leq p < \infty)$ (see Section 10; we give a new proof due to H. Lemberg [LE] which makes heavy use of model theoretic tools to achieve simplifications in the proof of this deep result); results due to S. Heinrich and P. Mankiewicz concerning the uniform homeomorphism of $\mathcal{L}_p$-spaces $(1 < p \leq \infty)$ (see Corollaries 11.4-.6); a new result due to the authors which gives a complete description of the asymptotic local geometry of the finite dimensional sequence spaces $\ell_p(n)$ as p and n tend to ∞ (see Theorem 8.3); examples showing the possibility of giving certain very explicit geometric characterizations of such classes of Banach spaces as the $C(K)$-spaces and the $\mathcal{L}_p$-spaces (see Theorem 6.3). In addition to these results there are many other standard applications of model theoretic methods which we do not have space to discuss, especially in the theory of operator ideals. (See the papers of S. Heinrich and A. Pietsch.) We also discuss some new and much easier proofs of important known results, for example: the easy proof by S. Heinrich and P. Mankiewicz of the theorem due to M. Ribe that any two uniformly homeomorphic Banach spaces are finitely represented in each other (see Theorem 11.2); the easy proof by S. A. Rakov of the 3-space theorem for super-reflexive spaces due to Enflo, Lindenstrauss and Pisier. (See Proposition 3.10.)

The detailed contents of this paper are laid out as follows: Section 1 contains a brief recollection of the framework for nonstandard analysis and of the most elementary facts about nonstandard hulls. Here we discuss the relationship between nonstandard analysis and the ultraproduct approach (briefly: they are equivalent.) Section 2 contains an assortment of specific examples of nonstandard hulls and arguments about them. This may be useful to the reader who has not encountered these ideas before; it does not lie in the main line of exposition. Section 3 contains a detailed discussion of the connection between nonstandard hulls and finite representability of

Banach spaces. Section 4 concerns Banach lattices; nonstandard hulls of lattices are examined in general and also the hulls of C(K)-spaces and L_p-spaces are shown to be spaces of the same type. Sections 5 and 8 develop an essential tool for comparing Banach spaces through their nonstandard hulls, namely the logic of positive bounded formulas. In particular, in Section 8 is introduced a new game-theoretic concept of equivalence between the local geometry of Banach spaces. Section 6 contains discussion of the classical Banach spaces, the C(K)- and L_p-spaces, as well as the closely related $\mathcal{L}_p$-spaces. In Section 7 we present one of the most important tools yet developed for applying nonstandard methods in Banach space theory, the local duality theorem due independently to K. D. Kürsten and J. Stern (see Theorem 7.1 and Corollary 7.2.) This result has as yet no standard counterpart. It underlies such applications as Heinrich's treatment of the UAP and Henson's localization result for L_p-spaces among others, for which no other proof is yet known than the original, model-theoretic one. Section 9 treats hyperfinite dimensional spaces, especially the nonstandard hulls of $\ell_p(n)$ spaces for infinite p and n. In Section 10 we give a version of H. Lemberg's proof of Krivine's theorem on block finite representability of the c_0 and ℓ_p bases. Section 11 presents some aspects of the treatment by S. Heinrich and P. Mankiewicz of problems concerning uniform homeomorphism of Banach spaces and other nonlinear problems. Finally in Section 12 we present a series of open problems which we feel are important for the future development of this subject.

Because of space limitations there are many interesting related topics which we have chosen not to discuss. Notably this includes the important work by A. Pietsch and S. Heinrich on operator ideals and finite representability of operators. We have also omitted work, such as that by S. Bellenot, A. Grainger, M. Wolff, ourselves and others, on more general aspects of topological vector space theory. Another important topic we do not threat here is operator theory, including work on the invariant subspace problem by A. Robinson and A. Bernstein.

We acknowledge gratefully a large number of helpful conversations, correspondence and preprints from many people, especially including S. Bellenot, J. Conroy, D. Cozart, S. Heinrich, W. A. J. Luxemburg, J. Stern, K. Stroyan and M. Wolff.

Section 1. Basics.

The basic system of nonstandard analysis that we use here is the one described and used by Stroyan and Luxemburg [STL]. (See also the book by Davis [DA].) this system was originally presented by Robinson and Zakon [RZ]. It is essentially equivalent to the type-theoretic formulation used by Robinson [ROB2], but it seems to be mathematically smoother to use than type theory.

In this system one begins with a collection of mathematical objects M, containing those objects which one wishes to study, such that M is a superstructure. This means that M consists of a set X of basic elements (urelements) together with all sets which can be obtained from X by some finite number of iterations of the power set operator. So M contains all subsets of X, all sets of these, etc., up through all the finite levels of the set-theoretical hierarchy over X.

In addition to M one has another such superstructure $*M$ together with an embedding $*$ from M into $*M$. If A is an object in M, then $*A$ is the nonstandard extension of A. Unless A is a finite set, $*A$ is strictly larger than $\{*a | a \in A\}$ and it is the extra, nonstandard elements of $*A$ on which the usefulness of this approach depends. Also $*M$ is taken to satisfy the crucial Transfer Principle: the $*$-transform of any formalized mathematical assertion true in M must be true in $*M$. For convenience we may assume that the $*$ mapping is the identity on X, so $X \subseteq *X$. (The corresponding relation will not be true for objects at higher levels in M.) We will always assume that X contains the real line $\mathbb{R}$. Thus $*r = r$ for all $r \in \mathbb{R}$ and $*\mathbb{R}$ is an extension ordered field of $\mathbb{R}$.

The terminology and notation from nonstandard analysis that we use is the same as that found in [STL]. In particular, suppose p and q are in $*\mathbb{R}$, the nonstandard extension of the real line $\mathbb{R}$. We say p is infinitesimal if $|p| \le r$ holds for every standard real number $r \in \mathbb{R} \subseteq *\mathbb{R}$; p is finite if $|p| \le r$ holds for some such r. If p is finite then there is a unique $r \in \mathbb{R}$ such that $p - r$ is infinitesimal. This number r is the standard part of such a finite p, written $r = st(p)$. When $p - q$ is infinitesimal we write $p \approx q$. An object S in $*M$ is said to be internal if there is some standard set A in M so that $S \in *A$. Otherwise S is external. This distinction between internal and external is of crucial importance in nonstandard analysis.

For us there are two other important properties that a nonstandard extension $*M$ can satisfy. If κ is an uncountable cardinal, then

$^*M$ is said to be <u>κ-saturated</u> [LUX 1] if given a family $\{S_i \mid i \in I\}$ of internal sets, all subsets of an internal set T, the hypotheses that card(I) $< \kappa$ and $\{S_i \mid i \in I\}$ has the finite intersection property always imply that the total intersection $\cap\{S_i \mid i \in I\}$ is not empty. Also $^*M$ is said to have the $\aleph_0$-isomorphism property [HEN1] if, for each first order language L with a finite number of relations, if A and B are elementarily equivalent structures for L whose domains, relations and functions are all internal, then A and B are isomorphic.

We will always assume that M contains $\mathbb{R}$ and that $^*M$ is at least $\aleph_1$-saturated. Usually we will also assume that it has the $\aleph_0$-isomorphism property, though this will normally be mentioned explicitly. (See the discussion in Section 5 of this property and its usefulness in the Banach space setting. See also Proposition 9.3.) We regard the structures M and $^*M$ solely as tools for proving results about standard objects of mathematics or for constructing such objects. The following result records the basic existence principle which insures that M and $^*M$ will always be available as needed. (See [LUX 1], [STL] and [HEN 1].)

<u>Proposition 1.1.</u> (a) If S is any set of mathematical objects, then there is a superstructure M which contains (a copy of) each member of S.

(b) If M is any superstructure and κ any uncountable cardinal, then there is a nonstandard extension $^*M$ of M such that $^*M$ is κ-saturated and has the $\aleph_0$-isomorphism property.

Note that because we take M to contain $\mathbb{R}$, then M will also contain an isometric copy of each separable Banach space. Also M is closed under formation of dual spaces, spaces of operators, etc. We adopt the usual terminology of Banach space theory, as presented in [LT 2,3] for example. Thus isomorphisms, isometries, projections and other operators are linear unless otherwise mentioned. Also all Banach spaces are taken over the real field. Nearly all the proofs given here are valid for complex coefficients as well as for the real field. In a few places (notably in the last part of Section 4) Banach algebra techniques are needed where lattice techniques are used here.

The particular system of nonstandard analysis used here can easily be replaced by the Internal Set Theory introduced by Nelson [NE] (see also [HRB]) or by ultraproduct techniques. These three approaches are completely equivalent in the sense that arguments in one system can systematically be translated into the others. In Banach space theory the ultraproduct approach has so far been widely used. (See [HEI 4] and the papers of J. Stern and S. Heinrich especially.) Strictly speaking, it is included in the nonstandard hull approach, in that $^*M$ can be constructed from an ultrapower extension of M, even satisfying the properties in Proposition 1.1(b). Moreover, in that case the nonstandard hulls constructed using $^*M$ are precisely the same as the Banach space ultraproducts formed using the same index set and ultrafilter that was used in constructing $^*M$. (See the introduction to [HEN 3].)

It is our opinion that the nonstandard analysis approach has definote conceptual advantages over the ultraproduct method (as it has been used in Banach space theory so far) in that nonstandard analysis has developed a natural collection of tools and concepts (such as the internal cardinality of a $*$-finite set) that are very useful and are not so easy to express or discover in the ultraproduct setting. We feel that the greater expressive power of the nonstandard analysis framework (whether the Robinson-Zakon system used here or Nelson's Internal Set Theory) makes it a superior tool for proof in mathematics and especially for discovery.

Now suppose we have fixed M and $^*M$ as above. Let V be an internal Banach space in $^*M$; this means that $V \in {}^*\mathcal{B}$ for some set $\mathcal{B}$ of Banach spaces such that $\mathcal{B} \in M$. (Sometimes we will refer to a Banach space in M as a standard Banach space, to emphasize that it is a Banach space in the usual sense.) The object V with its norm function $\|\cdot\|$ obeys the $*$-transform in $^*M$ of the usual definition. Thus $\|\cdot\|$ is a function into ${}^*\mathbb{R}^+$ which is homogeneous over ${}^*\mathbb{R}$, satisfies the usual triangle inequality and makes V complete in the internal sense, transformed to $^*M$.

An element p of V is finite if $\|p\|$ is finite in ${}^*\mathbb{R}$; it is infinitesimal if $\|p\| \approx 0$. We denote by $\mathrm{fin}(V)$ the set of finite elements of V and by $\mu(0)$ the set of infinitesimal elements. (The set $\mu(0)$ is also known as the monad of 0 in V). Both $\mathrm{fin}(V)$ and $\mu(0)$ are vector spaces over $\mathbb{R}$ (the standard real line) so we may define $\hat{V}$ to be the quotient vector space $\mathrm{fin}(V)/\mu(0)$ over $\mathbb{R}$. We always let

$$\pi\colon \mathrm{fin}(V) \to \hat{V}$$

denote the quotient linear mapping. Finally we define a norm on $\hat{V}$ by

$$\|x\| = st\|p\|$$

for all $p \in fin(V)$, $x = \pi(p) \in \hat{V}$. Under these definitions $\hat{V}$ becomes a Banach space over $\mathbb{R}$ in the usual sense. (The completeness of $\hat{V}$ comes because we have assumed $*M$ to be $\aleph_1$-saturated.) Suppose further that V is an internal Banach lattice under the partial ordering $\leq$. Then $\hat{V}$ also becomes a Banach lattice under an ordering $\leq$ defined as follows: given $x = \pi(p)$ and $y = \pi(q)$ in $\hat{V}$,

$$x \leq y \iff p \leq q + z$$

for some infinitesimal z in V.

For details about these constructions, see [STL] or [HMR 1] and [CZM].

An important special case occurs when V is $*E$ for some standard Banach space E in M. In that case we write $\hat{E}$ for $\hat{V}$ and refer to $\hat{E}$ as a <u>nonstandard hull of E</u>. We may embed E in $\hat{E}$ by sending each x in E to $\pi(*x)$. This is a linear isometric embedding; typically we regard E as a subspace of $\hat{E}$ using this mapping. If E is a Banach lattice then E becomes a sublattice of $\hat{E}$.

Also of importance is the case where V has $*$-finite dimension. That is, the internal dimension of V over $*\mathbb{R}$ is an integer in the sense of $*M$. This means that there is an internal sequence $x_1, x_2, \ldots, x_\omega$ ($\omega \in *\mathbb{N}$) in V which forms a basis for V in the sense that every element p of V can be uniquely represented as a $*$-finite sum

$$p = \sum_{j=1}^{\omega} \alpha_j x_j$$

where $\alpha_1, \ldots, \alpha_\omega$ is an internal sequence from $*\mathbb{R}$. In this case we call the nonstandard hull $\hat{V}$ a <u>hyperfinite dimensional</u> Banach space. These are generally very nonseparable, yet they share many of the formally expressible properties of finite dimensional spaces.

Suppose now that V and W are internal Banach spaces and that $S: V \to W$ is an internal linear operator (over $*\mathbb{R}$). Suppose also that S is bounded in the sense that there exists $K > 0$ in $*\mathbb{R}$ such that

$$\|S(p)\| \leq K\|p\|$$

for all $p \in V$. As in the standard setting we may define $\|S\|$ to be the inf in $^*\mathbb{R}$ of all such K. Since the set of all such K is internal, by the Transfer Principle, this inf will exist. Assume now that $\|S\|$ is finite in $^*\mathbb{R}$. It follows that S maps fin(V) into fin(W) and takes infinitesimals to infinitesimals. Hence we may define $\hat{S}: \hat{V} \to \hat{W}$ by

$$\hat{S}(x) = \pi(S(p))$$

where $x \in \hat{V}$, $x = \pi(p)$. Then $\hat{S}$ is linear over $\mathbb{R}$ and its norm is given by $\|\hat{S}\| = \mathrm{st}(\|S\|)$, as is easy to check.

In particular, we may take W to be $^*\mathbb{R}$ and consider internal linear functionals $\phi: V \to {}^*\mathbb{R}$. The above discussion shows that if $\|\phi\|$ is finite, then ϕ gives rise to a bounded linear functional $\hat{\phi}: \hat{V} \to \mathbb{R}$ whose norm is given by $\|\hat{\phi}\| = \mathrm{st}(\|\phi\|)$. This yields a linear isometric embedding from $\widehat{V'}$ into $(\hat{V})'$, where V' denotes the internal dual space of V. Put another way, this gives rise to a pairing between the nonstandard hulls $\hat{V}$ and $\widehat{V'}$ given by

$$\langle \pi(\phi), \pi(p) \rangle = \mathrm{st}(\langle \phi, p \rangle)$$

for each finite $p \in V$ and $\phi \in V'$. In particular, if E is a standard Banach space, then $E' \subseteq \widehat{E'} \subseteq (\hat{E})'$, so that we have a natural procedure for extending each bounded linear functional on E to one defined on all of $\hat{E}$. Moreover this extension procedure is norm-preserving and linear.

<u>Proposition 1.2.</u> Let V be an internal Banach space.

(a) If the internal dimension of V is a standard, finite integer m, then $\hat{V}$ has the same dimension over $\mathbb{R}$;

(b) otherwise, $\hat{V}$ is nonseparable.

Indeed, the unit ball of $\hat{V}$ contains a subset S such that for distinct x,y in S, $\|x-y\| \geq 1$ and the cardinality of S is $\geq \kappa$, where $^*M$ is a κ-saturated extension.

<u>Proof.</u> (a) Let $(p_1,\ldots,p_m)$ be an Auerbach basis for V with coordinate functionals $\phi_1,\ldots,\phi_m$ in V'. Thus each p_j and ϕ_j has norm 1 and $\phi_i(p_j) = 0$ if $i \neq j$, while $\phi_i(p_i) = 1$. Let $x_i = \pi(p_i)$ for $1 \leq i \leq m$. Using the linear functionals $\hat{\phi}_1,\ldots,\hat{\phi}_m$ we can show that $x_1,\ldots,x_m$ are linearly independent over $\mathbb{R}$. Moreover, if $p = \sum_{j=1}^{m} \alpha_j p_j$ is a finite element of V, then $|\alpha_j| = |\phi_j(p)| \leq$

$\|p\| \|\phi_j\| = \|p\|$ so that $\alpha_1,\ldots,\alpha_m$ are finite in $*\mathbb{R}$. From this it is easy to show that

$$\pi(p) = \sum_{j=1}^{m} st(\alpha_j)x_j .$$

Hence $x_1,\ldots,x_m$ is a basis for $\hat{V}$ over $\mathbb{R}$.

(b) If the internal dimension of V over $*\mathbb{R}$ is not in $*\mathbb{N}$, then there is an internal subset A of the unit ball of V such that for distinct $p,q \in A$, $\|p - q\| \geq 1$ and such that the internal cardinality of A is not finite. (For example, A could be an Auerbach basis for an internal subspace of V with internal dimension an infinite integer.) Then $S = \{\pi(p) | p \in A\}$ has the required properties.

<u>Remark.</u> When V is $*E$ for a standard Banach space E, Proposition 1.2(a) shows that $\hat{E}$ equals E if E is finite dimensional. Part (b) shows that if E is infinite dimensional and if $*M$ is κ-saturated, where κ is greater than the density character of E, then $E \subsetneq \hat{E}$.

<u>Lemma 1.3.</u> Let V be an internal Banach space. If $p_1,\ldots,p_n \in fin(V)$ and $\pi(p_1),\ldots,\pi(p_n)$ are linearly independent over $\mathbb{R}$, then $p_1,\ldots,p_n$ are $*$-linearly independent over $*\mathbb{R}$.

<u>Proof.</u> Suppose $\alpha_1,\ldots,\alpha_n \in *\mathbb{R}$ are not all 0 and $\sum_{j=1}^{n} \alpha_j p_j = 0$. Then we may assume that $\alpha_1 \neq 0$ and $|\alpha_1| \geq |\alpha_j|$ for all j. Dividing by α_1, we see that we may as well assume $\alpha_1 = 1$ and $|\alpha_j| \leq 1$ for all j. Then it is easy to show that since $\alpha_1,\ldots,\alpha_n$ are all finite,

$$\sum st(\alpha_j)\pi(p_j) = \pi(\sum \alpha_j p_j) = 0$$

which contradicts the independence of $\pi(p_1),\ldots,\pi(p_n)$.

<u>Lemma 1.4.</u> Let E be a standard Banach space and suppose $*M$ is κ-saturated where κ is greater than the density character of E. Then there exists a $*$-finite dimensional subspace V of $*E$ so that $E \subseteq \hat{V} \subseteq \hat{E}$.

<u>Proof.</u> Let $A \subseteq E$ be a dense subset of cardinality $< \kappa$. Since $*M$ is κ-saturated there exists a $*$-finite set $S \subseteq *E$ which satisfies

$$\{*a | a \in A\} \subseteq S.$$

Let V be the internal linear span of S, so V is a *-finite dimensional internal subspace of *E. Then $A \subseteq \hat{V}$ and $\hat{V}$ is closed, so $E \subseteq \hat{V}$ as desired.

Section 2. Examples.

In this section we introduce particular examples in order to illustrate the structure of nonstandard hulls and the arguments used in investigating them. The reader desiring to see the general theory may go on to the later sections and refer back to the present discussion as needed. In this section we state several results under the assumption that the nonstandard extension is of a particularly nice form, i.e., that it has the $\aleph_0$-isomorphism property. (See Section 5.) This property of the extension enables one to conclude the existence of isometries between various pairs of nonstandard hulls. However, even without this assumption such pairs of spaces are strongly related. (See the equivalence relation $\equiv_A$ described in Section 5.)

2.1. The spaces ℓ_p. The * transform of the set of all standard ℓ_p spaces $1 \leq p < \infty$ consists of the internal Banach spaces ${}^*\ell_p$ where $p \in {}^*\mathbb{R}$ and $p \geq 1$. We examine first ${}^*\ell_1$ and its nonstandard hull $\hat{\ell}_1$.

For each k in *N let e_k be the internal function from *N to ${}^*\mathbb{R}$ which assigns 1 to k and 0 to the other j in *N. Clearly each e_k belongs to $\mathrm{fin}({}^*\ell_1)$. We designate the image of e_k in $\hat{\ell}_1$ by $\hat{e}_k$ and identify it with the element of $\ell_1({}^*N)$ which assigns 1 to k and 0 to the other j. More generally, $\ell_1({}^*N)$ may be considered as a subspace of $\hat{\ell}_1$. To see this let $h = \sum_{j=1}^{\infty} c_j \hat{e}_{k_j}$ be an element of $\ell_1({}^*N)$ where we assume $k_i \neq k_j$ for $i \neq j$. Using $\aleph_1$-saturation, we may define an internal function f in ${}^*\ell_1$ with *-finite support such that $f(k_j) = c_j$ for finite j and $\|f\| =_1 \|h\|$. Clearly $\pi(\sum_{j=1}^{n} c_j e_{k_j}) = \sum_{j=1}^{n} c_j \hat{e}_{k_j} \to h$, i.e., we may identify $\pi(f)$ and h. Not only is $\ell_1({}^*N)$ a subspace $\hat{\ell}_1$, but it is easy to verify that $\ell_1({}^*N)$ is the projection band in $\hat{\ell}_1$ generated by the atoms of $\hat{\ell}_1$ and contains, in turn, the original sequence space ℓ_1 as a projection band.

However there also is a nonatomic part to $\hat{\ell}_1$. Let θ be an infinite positive integer and let f be $1/\theta$ times the characteristic function of the interval $\{1,2,\ldots,\theta\}$. Clearly $f \in \mathrm{fin}({}^*\ell_1)$ and both f and $\pi(f)$ are positive of norm 1. Moreover if g is positive and $\pi(g)$ is in $\ell_1({}^*N)$, it is easy to see that $\inf(f,g) \approx 0$, so $\pi(f) \in \ell_1({}^*N)^{\perp}$. Indeed $\ell_1({}^*N)^{\perp}$ is a non-atomic abstract L_1-

space and so by the Kakatani representation theory may be identified with $L_1(\mu)$ for some measure μ. Hence $\hat{\ell}_1$ has the form $\ell_1({}^*N) \oplus L_1(\mu)$. It is shown in [HEN 3] that if $^*M$ has the $\aleph_0$-isomorphism property then $L_1(\mu)$ is order isometric to the nonstandard hull of $L_1[0,1]$.

Similarly if p is standard, $1 < p < \infty$, then $\hat{\ell}_p$ may be represented in the form $\ell_p({}^*N) \oplus L_p(\mu)$ where $L_p(\mu)$ has no atoms. In particular $\hat{\ell}_2$ is again a Hilbert space with inner product $(\pi(f), \pi(g)) = \mathrm{st}\ {}^*(f,g)$.

Now suppose p is a finite element of $^*\mathbb{R}$ with $1 \leq p$ and $\mathrm{st}(p) = r$. Then if $0 \leq f,g \in \mathrm{fin}({}^*\ell_p)$ and $\inf(f,g) = 0$, we have $\|f + g\|^p = \|f\|^p + \|g\|^p$. It follows that $\|\pi(f) + \pi(g)\|^r = \|\pi(f)\|^r + \|\pi(g)\|^r$. If $0 \leq u,v \in \hat{\ell}_p$ and $\inf(u,v) = 0$, we can find $0 \leq f,g \in \mathrm{fin}({}^*\ell_p)$ with $\inf(f,g) = 0$ and $\pi(f) = u$, $\pi(g) = v$. It follows that $\hat{\ell}_p$ is an abstract L_r-space and can be written $\ell_r({}^*N) \oplus L_r(\mu)$ where $L_r(\mu)$ has no atoms. In Section 6 we show that if $^*M$ has the $\aleph_0$-isomorphism property, then $\hat{\ell}_p$ is order isometric to $\hat{\ell}_r$.

Let p be a positive infinite element in $^*\mathbb{R}$. Then the fact that $0 \leq f,g \in \mathrm{fin}({}^*\ell_p)$ and $\inf(f,g) = 0$ imply $\|f + g\|^p = \|f\|^p + \|g\|^p$ translates into the fact that $0 \leq u,v$ in $\hat{\ell}_p$ and $\inf(u,v) = 0$ imply $\|u + v\| = \max(\|u\|, \|v\|)$, i.e., $\hat{\ell}_p$ is an abstract M-space. It is easy to verify that the elements $\pi(e_k)$ are exactly the positive norm-one atoms in $\hat{\ell}_p$, but $\hat{\ell}_p$ is not completely atomic i.e., not every positive element is the supremum of the positive atoms it dominates. For example, let θ be a positive infinite integer and pick ϕ positive infinite so that $\phi/\theta^p \approx 1$. If $f \in {}^*\ell_p$ is defined by

$$f(k) = \begin{cases} \frac{1}{\theta} & 1 \leq k \leq \phi \\ 0 & k > \phi, \end{cases}$$

then $\pi(f)$ is a positive norm-one element of $\hat{\ell}_p$ and $\inf(\pi(f), \pi(e_k)) = 0$ for all k in $^*N$. We show below that $\hat{\ell}_p$ is not isometric as a Banach space to $\hat{\ell}_\infty$. If $^*M$ has the $\aleph_0$-isomorphism property, then the spaces $\hat{\ell}_p$ for infinite p are shown to be pairwise order isometric in Section 9.

2.2. <u>The spaces</u> $\hat{\ell}_\infty$ <u>and</u> $\hat{c}_0$. For $f,g \in \mathrm{fin}({}^*\ell_\infty)$ we have $\pi(f) = \pi(g)$ if and only if $f(k) \approx g(k)$ for all k in $^*N$. It follows that $\hat{\ell}_\infty$ may be order isometrically embedded in $\ell_\infty({}^*N)$ by identifying $\pi(f)$ with the function h defined by $h(k) = \mathrm{st}\ f(k)$ for all k in $^*N$. Under this identification, $\hat{\ell}_\infty$ is a proper subspace of $\ell_\infty({}^*N)$

since, for example, the characteristic function of N does not belong to $\hat{\ell}_\infty$. (If $f \in \text{fin}(*\ell_\infty)$ and $f(k) \approx 1$ for all k in N, then $\text{st}(f(k)) > 0$ for some infinite k.) Indeed, it is easy to check that $\hat{\ell}_\infty$ is the closure in $\ell_\infty(*N)$ of the span of the characteristic functions of internal subsets of *N.

If $f \in \text{fin}(*c_0)$ then for some $\theta \in *N$ we have $f(k) \approx 0$ for $k > \theta$. Since the characteristic function of every *-finite set belongs to $\text{fin}(*c_0)$, it follows that, under the identification described above, $\hat{c}_0$ is the closure in $\ell_\infty(*N)$ of the span of the characteristic functions of *-finite sets. Clearly both $\hat{\ell}_\infty$ and $\hat{c}_0$ are completely atomic abstract M-spaces. While $\hat{\ell}_\infty$ has a strong order unit (the characteristic function of *N), $\hat{c}_0$ does not have a weak order unit. (By the Transfer Principle, if $0 \leq f \in *c_0$ with $\|f\| = 1$ and α is a positive infinitesimal, there exists $0 \leq g$ in $*c_0$ of norm 1 such that both $\|f + g\|$ and $\|f - g\|$ are between $1 - \alpha$ and $1 + \alpha$. It follows that for every positive element u of norm 1 in $\hat{c}_0$ there exists a positive element v of norm 1 in $\hat{c}_0$ with $\inf(u,v) = 0$.)

Since the space $\hat{\ell}_p$ for infinite p is not completely atomic, it follows that $\hat{\ell}_p$ is not order isometric to $\hat{c}_0$ and since $\hat{\ell}_p$ does not have a strong order unit $\hat{\ell}_p$ is not isometric as a Banach space to $\hat{\ell}_\infty$. (The space $\hat{\ell}_\infty$ satisfies the property that there exists u of norm 1 such that if v has norm 1 then either $\|u + v\| = 2$ or $\|u - v\| = 2$, but this property does not hold for $\hat{\ell}_p$.) We show in Section 9 that if p is infinite then $\hat{\ell}_p$ is not isometric as a Banach space to $\hat{c}_0$.

2.3. $\hat{\ell}_\infty$ __is not the dual space of__ $\hat{\ell}_1$. As noted in Section 1, $\hat{\ell}_\infty$ embeds naturally as a subspace of $\widehat{\ell}_1'$. Indeed the pairing between the spaces is given by

$$\langle \pi(f), \pi(g) \rangle = \text{st}^*\langle f,g \rangle \quad \text{for } f \in \text{fin}(*\ell_1) \text{ and } g \in \text{fin}(*\ell_\infty).$$

We show that $\hat{\ell}_\infty$ is not all of $\widehat{\ell}_1'$. Define a linear functional ϕ on $\hat{\ell}_1$ by

$$\phi(\pi(f)) = \lim_{n\to\infty} \text{st} \sum_{k=1}^{n} f(k) .$$

Here n varies over the standard positive integers. It is easy to verify that ϕ is a well-defined positive linear functional on $\hat{\ell}_1$ of norm one. If ϕ is $\pi(f)$ for some $f \in fin(*\ell_\infty)$, then since $\phi(\hat{e}_k) = 1$ for all finite k, we have $f(k) \approx 1$ for all finite k. Hence $f(\theta) \approx 1$ for some infinite θ which is a contradiction since $\phi(\hat{e}_k) = 0$ for all infinite integers k. In Section 3 we describe for what $\hat{E}$ we can identify $\hat{E}'$ and $\widehat{E'}$ with each other (E must be super-reflexive) and in Section 7 we discuss a deep result obtained by both Stern and Kürsten describing in general the relation between $\hat{E}'$ and $\widehat{E'}$.

2.4. $\hat{\ell}_\infty$ __is not injective__. Recall that ℓ_∞ is an injective space, i.e., for any Banach space X containing ℓ_∞ there is a bounded linear projection of X onto ℓ_∞. (See [LT 2].) We prove that $\hat{\ell}_\infty$ is not injective by showing that there does not exist a bounded linear projection of $\ell_\infty(*N)$ onto $\hat{\ell}_\infty$. The argument is based on the fact that for any infinite set S there does not exist a bounded linear projection of $\ell_\infty(S)$ onto $c_0(S)$ [LT 2] and the observation made in Section 1 that infinite internal sets are uncountable.

Assume that there exists a bounded linear projection P of $\ell_\infty(*N)$ onto $\hat{\ell}_\infty$. Let S be a countable subset of $*N$ and T an infinite internal subset of $*N$. Since T is uncountable we may pick a countably infinite subset W of T such that $W \cap S = \emptyset$. Let Q be the canonical projection of $\ell_\infty(*N)$ onto $\ell_\infty(W)$ obtained by restricting h in $\ell_\infty(*N)$ to W. Since $c_0(W) \subseteq \hat{\ell}_\infty$, we cannot have that the range of QP restricted to $\ell_\infty(W)$ is contained in $c_0(W)$. Otherwise QP restricted to $\ell_\infty(W)$ would be a bounded projection of $\ell_\infty(W)$ onto $c_0(W)$. Thus there exists a standard positive α and a function h of norm 1 in $\ell^\infty(*N)$ with support in W such that $\{k \in W: (Ph)(k) > 0\}$ is infinite. Since $Ph \in \hat{\ell}_\infty$, there is an internal subset T_1 of T such that $T_1 \cap W$ is infinite and $(Ph)(k) \geq \alpha$ for all $k \in T_1$. Let $\beta(S,T)$ be the supremum of the real numbers α so that there exists a norm-one function h in $\ell_\infty(*N)$ with countable support contained in $T \sim S$ and an infinite internal subset T' of T such that $(Ph)(k) \geq \alpha$ for all k in T'. Clearly $0 < \beta(S,T) \leq \|P\|$.

Starting with $T_1 = *N$ and $S_0 = \emptyset$, we inductively construct sequences $\{h_n\}$, $\{S_n\}$, and $\{T_n\}$ $n = 1,2,\ldots$ so that

a) for $n \geq 1$, T_n is an infinite internal set and $T_{n+1} \subseteq T_n$;

b) for $n \geq 1$, h_n is a norm-one element of $\ell_\infty(*N)$ with support S_n where S_n is a countably infinite subset of T_n;

c) for $n > m \geq 1$, $S_n \cap S_m = \emptyset$;

d) for $n \geq 1$, $(Ph_n)(k) \geq (1/2)\beta(S_1 \cup \ldots \cup S_{n-1}, T_n)$ for all

k in T_{n+1}.

Clearly the sequence $\beta_n = \beta(S_1 \cup \ldots \cup S_{n-1}, T_n)$ is monotone decreasing. Suppose $\beta_n \downarrow 0$. In this case let J be a set of elements k_j of $*\mathbb{N}$ such that $k_j \in S_j$ for $j = 1,2,\ldots$ and let R be the canonical projection of $\ell_\infty(*\mathbb{N})$ onto $\ell_\infty(J)$. Now if g has norm 1 and suppose in J, define $g_n = g - \sum_{j=1}^{n} g(k)\hat{e}_{k_j}$. Then $\|g_n\| \leq 2$ and g_n has support in $T_{n+1} \sim S_1 \cup \ldots \cup S_n$. It follows that $|(Pg_n)(k_j)| = |(Pg)(k_j)| \leq 2\,\beta_{n+1}$ for all but a finite number of j. Thus $RP(g) \in c_0(J)$ for all $g \in \ell_\infty(J)$, which implies RP defines a bounded linear projection of $\ell_\infty(J)$ onto $c_0(J)$, a contradiction.

Thus for some real $\beta > 0$, we have $\beta_n \geq \beta$ for all n. Then the function

$$w_n = \sum_{j=1}^{n} h_j$$

has norm 1 since the h_j's have disjoint support, but $P\,w_n(k) \geq \frac{n\beta}{2}$ for all k in T_{n+1}, which contradicts the assumtion that P is bounded.

2.5. Examples of hyperfinite-dimensional spaces. The nonstandard hulls $\hat{\ell}_p(n)$ of the *-finite dimensional spaces $*\ell_p(n)$ for $1 \leq p$ in $*\mathbb{R}$ and n in $*\mathbb{N}$ provide a rich assortment of hyperfinite-dimensional spaces. If p and n are finite, it is easy to see that $\hat{\ell}_p(n)$ is order isometric to $\ell_r(n)$ where $r = st(p)$. If n is finite and p is infinite then $\hat{\ell}_p(n)$ is order isometric to $\ell_\infty(n)$. (For any x in $\mathbb{R}^n$, $\lim_{p\to\infty} \|x\|_p = \|x\|_\infty$.)

If p is finite and n is infinite, then it is easy to verify that $\hat{\ell}_p(n)$ is an abstract L_r-space, where $r = st\,p$, and assuming the $\aleph_0$-isomorphism property the space $\hat{\ell}_p(n)$ is order isometric to $\hat{\ell}_r$. If both p and n are infinite, there exists a continuum of non-isometric spaces $\hat{\ell}_p(n)$ depending on the relative size of p and n. In particular if n is large with respect to p (if $n^{1/p}$ is infinite), then, assuming the $\aleph_0$-isomorphism property, $\hat{\ell}_p(n)$ and $\hat{\ell}_p$ are order isometric. (See Section 9.)

2.6. A hyperfinite-dimensional Banach space whose unit ball does not contain an extreme point.

We construct a sequence of finite-dimensional subspaces F_n of ℓ_∞ with each F_n having the property that for each f of norm 1 there exists g of norm 1 such that both $\|f + g\|$ and $\|f - g\|$

lie between $1 - 1/n$ and $1 + 1/n$. Then if θ is an infinite positive integer, it is easy to verify that the hyperfinite-dimensional Banach space $\widehat{{}^*F_\theta}$ has the property that for each h of norm 1 there exists w of norm 1 so that $\|h + w\| = \|h - w\| = 1$. Since $h = \frac{1}{2}(h + w) + \frac{1}{2}(h - w)$, it follows that the unit ball of $\widehat{{}^*F_\theta}$ has no extreme points.

Let n be a fixed standard positive integer and let $\{f_j : 1 \le j \le n + 1\}$ be elements of ℓ_∞ of the form

$$f_j = (\underbrace{0,\ldots,0,1,0,\ldots,0}_{n+1 \text{ places}},\underbrace{\pm 1/n,\pm 1/n,\ldots,\pm 1/n}_{2^{n+1} \text{ places}},0,0,\ldots)$$

such that:

(i) for each j, $f_j(j) = 1$ and $f_j(k) = 0$ for $k \ne j$, $k = 1,2,\ldots,n+1$.

(ii) for each finite sequence $\varepsilon_1,\varepsilon_2,\ldots,\varepsilon_{n+1}$ of $+1$, -1 there is a k such that $f_j(k) = \varepsilon_j \frac{1}{n}$ for all j.

Let $F_n = \operatorname{span}\{f_1,f_2,\ldots,f_{n+1}\}$. If x is in F_n with $\|x\| = 1$, then $f = \sum_{j=1}^{n+1} c_j f_j$ where $|c_j| \le 1$ for all j. Also $|c_j| \ge 1/n$ for at most n of the j's, since there exists k such that $f(k) = \frac{1}{n}\sum_{j=1}^{n+1} |c_j|$. Pick j_0 so that $|c_{j_0}| \le 1/n$. Then

$$1 - 1/n \le 1 - |c_{j_0}| \le |(f + f_{j_0})(j_0)| \le 1 + |c_{j_0}| \le 1 + 1/n$$

and if $k \ne j_0$ then $|(f + f_{j_0})(k)| \le |f(k)| + 1/n \le 1 + 1/n$. It follows that $1 - 1/n \le \|f + f_{j_0}\| \le 1 + 1/n$ and similarly $1 - 1/n \le \|f - f_{j_0}\| \le 1 + 1/n$, i.e., we may take $g = f_{j_0}$.

Section 3. Basic Geometry.

Although nonstandard hulls may be quite large (see Proposition 1.2) their structure is determined to a considerable degree by their finite dimensional subspaces. We need the following definitions:

Definition 3.1. Let E and F be standard Banach spaces and let $\lambda \ge 1$.

(i) The space E is said to be finitely λ-representable in F if for each finite-dimensional subspace G of E and each positive real number ε, there exists a linear transformation T of G into F such that

$$\|x\| \le \|Tx\| \le (\lambda+\varepsilon)\|x\| \quad \text{for all } x \text{ in } G.$$

If E is finitely 1-representable in F, we say simply that E is <u>finitely</u> <u>representable</u> in F.

(ii) A linear transformation T of E into F is called a λ-<u>embedding</u> if

$$\|x\| \le \|Tx\| \le \lambda\|x\| \quad \text{for all } x \text{ in } E;$$

it is a λ-isomorphism if, in addition, T maps E onto F.

Our first result states that a nonstandard hull (even of an internal Banach space) contains as subspaces all the (small) spaces which are finitely represented in it. (See the remark following the proof for the most general form of this result.)

<u>Theorem 3.2.</u> Let E be a separable Banach space, X an internal Banach space and $\lambda \ge 1$ a standard number. Then E is finitely λ-representable in $\hat{X}$ if and only if there is a λ-embedding of E into $\hat{X}$. In particular, E is finitely representable in $\hat{X}$ if and only if E is isometrically embeddable in $\hat{X}$.

<u>Proof.</u> Clearly there is only one direction to establish. So assume E is finitely λ-representable in $\hat{X}$. Since E is separable there is an increasing family $\{G_n: n = 1,2,\ldots\}$ of finite dimensional subspaces of E such that the dimension of G_n is n and $\bigcup_{n\in N} G_n$ is dense in E. Fix n and let $\{e_1,\ldots,e_n\}$ be a basis for G_n. By assumption there exists a linear transformation T_n of G_n into $\hat{X}$ such that $\|x\| \le \|T_n x\| \le (\lambda + 1/2n)\|x\|$ for all x in G_n. Pick $p_1,p_2,\ldots,p_n$ in X such that $\pi(p_i) = T_n(e_i)$ for $i = 1,2,\ldots,n$. Then by Lemma 1.3 $\{p_1,p_2,\ldots,p_n\}$ is *-linearly independent in X and the mapping Q_n of $^*G_n$ onto *-span$\{p_1,\ldots,p_n\}$ given by

$$Q_n\left(\sum_{k=1}^{n} \lambda_k \, {}^*e_k\right) = \sum_{k=1}^{n} \lambda_k p_k$$

is an internal, linear 1-1 mapping of $^*G_n$ into X satisfying

$$(\#) \qquad (\lambda - 1/n)\|q\| \le \|Q_n(q)\| \le (\lambda + 1/n)\|q\|$$

$$\text{for all } q \text{ in } {}^*G_n \text{ with } \|q\| = 1.$$

Now the set of n in *N such that there exists an internal,

linear, 1-1 mapping Q_n of $^*G_n$ into X satisfying (#) is an internal set containing all finite positive integers. Hence it must contain an infinite positive integer θ. Now $\hat{Q}_\theta$ defined by

$$\hat{Q}_\theta(\pi(q)) = \pi(Q_\theta(q))$$

defines a λ-embedding of $\hat{G}_\theta$ into $\hat{X}$. Since $\bigcup_{n\in N} G_n$ is naturally contained in $\hat{G}_\theta$, the extension to E of the restriction of $\hat{Q}_\theta$ to $\bigcup_{n\in N} G_n$ gives the desired λ-embedding of E into $\hat{X}$.

Remark. When the nonstandard extension $^*M$ is κ-saturated, it can be shown that if E has density character less than κ, then E is finitely λ-representable in $\hat{X}$ if and only if there is a λ-embedding of E into $\hat{X}$.

For nonstandard hulls of standard Banach spaces more can be said.

Corollary 3.3. Assume $^*M$ is κ-saturated and let E,F be standard Banach spaces. Suppose F has density character $< \kappa$. Let $\lambda \geq 1$ be a standard number. Under these hypotheses, the following conditions are equivalent:

1) For each $\varepsilon > 0$, F is λ-finitely represented in E.
2) F is λ-finitely represented in $\hat{E}$.
3) F is λ-embeddable in $\hat{E}$.

Proof. The equivalence of (2) and (3) is part of Theorem 3.2. The implication from (1) to (2) is trivial since $E \subseteq \hat{E}$. The implication from (2) to (1) follows from the fact that $\hat{E}$ is finitely represented in E, which we prove below as part of Proposition 3.8.

Corollary 3.3 gives the most useful basic connection between a Banach space and its nonstandard hulls. Note that it has the consequence that if F is λ-finitely represented in E then F is λ-embeddable in some nonstandard hull of E, since we are free to choose $^*M$ to be κ-saturated for a large enough κ. This connection is strengthened some in Proposition 3.8 and it is given in its strongest form in Proposition 5.4.

Next we examine the question of which nonstandard hulls are reflexive. In this discussion the following characterizations of reflexive Banach spaces by R. C. James [JA] are useful.

Theorem 3.4. For a Banach space E the following are equivalent:

(i) E is reflexive.

(ii) Every separable subspace of E is reflexive.

(iii) For every linear functional y in E' there exists x in E such that $\|x\| = 1$ and $\langle x,y\rangle = \|y\|$.

Recall that a Banach space E is said to be super-reflexive if F finitely representable in E implies F is reflexive. It follows from Theorem 3.3 that E is super-reflexive if and only if every separable Banach space finitely representable in E is reflexive. Clearly every super-reflexive Banach space is reflexive, but it is easy to construct examples of reflexive spaces which are not super-reflexive. However for nonstandard hulls we have the following:

Corollary 3.5. A nonstandard hull $\hat{X}$ is reflexive if and only if it is super-reflexive.

Proof. One direction is trivial, so assume $\hat{X}$ is reflexive. If E is a separable Banach space which is finitely representable in $\hat{X}$, then by Theorem 3.2 E is isometrically embeddable in $\hat{X}$. Since $\hat{X}$ is reflexive, it follows that E also is reflexive. Hence $\hat{X}$ is super-reflexive.

Next we introduce an extension of the concept of finite representability which plays a central role in this section and in Section 7.

Definition 3.6. Let E be a Banach space and let G be a closed subspace. Then G is said to be a reflecting subspace of E if for every finite dimensional subspace F of E and every $\varepsilon > 0$ there exists a 1-1 linear transformation T of F into G such that:

(i) $\|T\| \le 1 + \varepsilon$ and $\|T^{-1}\| \le 1 + \varepsilon$

(ii) $Tx = x$ if $x \in G \cap F$.

The best-known class of examples of reflecting subspaces is described by the Principle of Local Reflexivity which states in part that a Banach space is a reflecting subspace of its second dual. (See [LT2]; this result was proved by J. Lindenstrauss and H. P. Rosenthal.)

Proposition 3.7. (Principle of Local Reflexivity) Let E be a Banach space. For each finite-dimensional subspace F of E", each finite subset $\{y_1,\ldots,y_n\}$ of E', and each $\varepsilon > 0$ there is a 1-1 linear transformation T of F into E such that

(i) $\|T\| \le 1+\varepsilon$ and $\|T^{-1}\| \le 1+\varepsilon$.

(ii) $Tx = x$ if $x \in F \cap E$.

(iii) $\langle y_i, x\rangle = \langle Tx, y_i\rangle$ for all x in F and $i = 1,2,\ldots,n$.

We return to the Principle of Local Reflexivity later in this section. The next result shows that E and $\hat{E}$ obey a similar property. In particular it yields that E is a reflecting subspace of $\hat{E}$. Recall that E' is considered a subspace of $\widehat{E'}$ and thus also a subspace of $(\hat{E})'$.

Proposition 3.8. Let E be a Banach space, let F be a finite dimensional subspace of $\hat{E}$, and let $\{y_1,\ldots,y_n\}$ be a finite subset of E'. Then for every $\varepsilon > 0$ there is a 1-1 linear transformation T of F into E such that:

(i) $(1-\varepsilon)\|x\| \le \|Tx\| \le (1+\varepsilon)\|x\|$ all x in F,

(ii) $Tx = x$ for all x in $E \cap F$,

(iii) $\langle x, y_i\rangle = \langle Tx, y_i\rangle$ for all x in F and $i = 1,2,\ldots,m$.

Proof. Let F, $\{y_1,\ldots,y_m\}$, and ε be as in the statement of the proposition. Clearly there exists a finite dimensional Banach space H in M which extends $F \cap E$ and an isometry Φ of H onto F which leaves $F \cap E$ invariant. Define linear functionals z_i, $i = 1,\ldots,m$ on H by $\langle x, z_i\rangle = \langle \Phi(x), y_i\rangle$ for all x in H. Let $\{e_1,\ldots,e_k\}$ be a basis for $F \cap E$ and let $\{e_1,\ldots,e_n\}$ be an extension to a basis for H. Pick elements $p_{k+1},\ldots,p_n$ of $\mathrm{fin}({}^*E)$ such that $\pi(p_j) = \Phi(e_j)$ for $j = k+1,\ldots,n$ and $\langle p_j, {}^*y_i\rangle = \langle \Phi(e_j), y_i\rangle$ for $j = k+1,\ldots,n$ and $i = 1,\ldots,m$. Then by Lemma 1.3 $\{{}^*e_1, {}^*e_2, \ldots, {}^*e_k, p_{k+1},\ldots,p_n\}$ is a *-linearly independent subset of $^*E$ which spans an internal subspace W of dimension n.

Define a 1-1 *-linear transformation Q of $^*H$ onto W by $Q(\sum_{j=1}^{n} \lambda_j {}^*e_j) = \sum_{j=1}^{k} \lambda_j {}^*e_j + \sum_{j=k+1}^{n} \lambda_j p_j$. If $p = \sum_{j=1}^{n} \lambda_j {}^*e_j$ has norm one in $^*H$, then $x = \sum_{j=1}^{n} \mathrm{st}(\lambda_j) e_j$ has norm one in H and $\pi(Q(p)) = \Phi(x)$ is a norm-one element of F. Thus $\|Q(p)\| \approx 1$. Moreover Q leaves $^*(E \cap F)$ invariant and

$$\langle Q(^*e_j), z_i\rangle = \langle \Phi(e_j), y_i\rangle$$

for $i = 1,\dots,m$ and $j = 1,\dots,n$. Thus by the Transfer Principle, if $\varepsilon > 0$ there exists a 1-1 linear transformation R of H into E such that

(i) $(1-\varepsilon)\|x\| \le \|Rx\| \le (1+\varepsilon)\|x\|$ all x in H

(ii) $Rx = x$ for all x in $F \cap E$,

(iii) $\langle e_j, z_i\rangle = \langle \Phi(e_j), y_i\rangle$ $i = 1,\dots,m$ and $j = 1,\dots,n$.

It is easy to verify that $R\Phi^{-1}$ is the desired transformation T.

Corollary 3.9. For any Banach space E the following are equivalent:

(i) E is super-reflexive,

(ii) $\hat{E}$ is reflexive,

(iii) $\hat{E}$ is super-reflexive.

Proof. The equivalence of (ii) and (iii) is given in Corollary 3.5. If $\hat{E}$ is super-reflexive, so is E since it is a subspace. Finally if E is super-reflexive, $\hat{E}$ is reflexive by Proposition 3.8.

The following observation of Rakov [HEI 4] provides a nice application of these ideas. Recall that reflexivity is a three-space property, i.e., if E is a Banach space and F is a closed subspace such that both F and E/F are reflexive, then E is reflexive [DS]. Enflo, Lindenstrauss and Pisier showed, using martingale inequalities, that super-reflexivity is also a three-space property [ELP]. Corollary 3.9 provides a simple proof of this result.

Proposition 3.10. If E is a Banach space, F is a closed subspace and if E and E/F are super-reflexive, then E is super-reflexive.

Proof. It is an easy calculation that $(\frac{E}{F})\hat{}$ is canonically isometric to $\hat{E}/\hat{F}$. Then if F and E/F are super-reflexive, Corollary 3.9 implies both $\hat{F}$ and $\hat{E}/\hat{F}$ are reflexive. By the classical result $\hat{E}$ is reflexive and so, by Corollary 3.9 again, E is super-reflexive.

We return to the relation between $\widehat{X'}$ and $\hat{X}'$.

Proposition 3.11. For any internal Banach space X the following are equivalent:

(i) $\hat{X}$ is reflexive,

(ii) $\hat{X}$ is super-reflexive,

(iii) $\widehat{X'} = \hat{X}'$.

Proof. The equivalence of (i) and (ii) is the content of Corollary 3.5. Now let $\hat{X}$ be reflexive and assume $\widehat{X'}$ is a proper closed subspace of $\hat{X}'$. Then there exists $x \in X$ such that $\|x\| = 1$ and $\langle x,y\rangle = 0$ for all $y \in \widehat{X'}$. But if $\pi(p) = x$ and $\|p\| = 1$, there exists $q \in {}^*X'$ with $\|q\| = 1$ such that $\langle p,q\rangle = 1$. (Use the Transfer Principle with the Hahn-Banach Theorem.) It follows that $\pi(q)$ is an element of $\widehat{X'}$ such that $\langle x,\pi(q)\rangle = 1$, a contradiction. Thus $\widehat{X'} = \hat{X}'$.

Now let $\widehat{X'} = \hat{X}'$. In order to show $\hat{X}$ is reflexive, it is enough to show that every linear functional in $\hat{X}'$ achieves its norm on the unit ball of $\hat{X}$. So let ϕ be a norm-one element of $\hat{X}' = \widehat{X'}$. Then there exists q in ${}^*X'$ with $\|q\| = 1$ and $\pi(q) = \phi$. But by the Transfer Principle for every positive α in ${}^*\mathbb{R}$ there exists p in ${}^*X$ with $\|p\| = 1$ such that $\langle p,q\rangle \geq 1 - \alpha$. Letting α be a positive infinitesimal, we obtain p such that $\|\pi(p)\| = 1$ and $\langle\pi(p),\phi\rangle = \mathrm{st}\langle p,q\rangle = 1$. Thus $\hat{X}$ is reflexive.

Next we show, subject to certain cardinality restrictions, that if E is isometric to a subspace of $\hat{X}$ and if E is a reflecting subspace of F then the isometry may be extended to an embedding of F into $\hat{X}$. In Proposition 3.13 we apply this to case where $F = E''$.

Proposition 3.12. Let X be an internal Banach space, let E be a standard Banach space and let Φ an isometry of E into $\hat{X}$. If ${}^*M$ is κ-saturated and E is a reflecting subspace of a Banach space F where card $F < \kappa$, then there is an extension Ψ of Φ to an isometric embedding of F into $\hat{X}$.

Proof. Let $\mathcal{F}$ be the collection of all finite dimensional subspaces of F. For each positive standard integer n and each $G \in \mathcal{F}$ define $A(n,G)$ to be the internal set of all *-linear functions P from a *-finite dimensional subspace of ${}^*F$ into ${}^*X$ such that

(i) the domain of P contains ${}^*G$,

(ii) for all q in the domain of P, $(1 - 1/n)\|q\| \leq \|Pq\| \leq (1 + 1/n)\|q\|$ and

(iii) for all q in ${}^*(E \cap G)$, $P(q) = {}^*\Phi(q)$.

Assume now that P is contained in the intersection of the sets $A(n,G)$ for $n \geq 1$ and $G \in \mathcal{F}$. Then if W is the domain of P, we have $F \subseteq \hat{W} \subseteq \hat{F}$ by (i). Moreover the linear mapping Ψ defined on F by $\Psi(x) = \pi(P(*x))$ is an isometry of F into $\hat{X}$ by (ii) and Ψ restricted to E is Φ by (iii).

Thus we only need to show that the intersection of the sets $A(n,G)$ is nonempty. Since the collection has fewer than κ-members and $*M$ is κ-saturated, it is sufficient to show that the collection has the finite intersection property. Indeed since any finite family of finite dimensional subspaces of F is again contained in a finite dimensional subspace of F, it is enough to show that each set $A(n,G)$ is nonempty.

Let n be a positive integer and $G \in \mathcal{F}$. Since E is a reflecting subspace of F there exists a 1-1 linear mapping Q of G into E such that

(a) $(1 - 1/2n)\|x\| \leq \|Qx\| \leq (1 + 1/2n)\|x\|$ all x in G

and (b) $Qx = x$ for all x in $G \cap E$.

If $\{z_1, z_2, \ldots, z_m\}$ is a basis for G, pick $\{p_1, \ldots, p_m\}$ in X such that $\Phi(Q(z_i)) = \pi(p_i)$ for $i = 1,2,\ldots,m$. Then we may define an internal linear transformation P from $*G$ into $*X$ by

$$P\left(\sum_{i=1}^{m} \lambda_i *z_i\right) = \sum_{i=1}^{m} \lambda_i p_i$$

for each sequence $\lambda_1, \ldots, \lambda_m$ in $*\mathbb{R}$. Now if $\left\|\sum_{i=1}^{m} \lambda_i *z_i\right\| = 1$, then each λ_i is finite and $P\left(\sum_{i=1}^{m} \lambda_i *z_i\right) \approx *\Phi\left(\sum_{i=1}^{m} st(\lambda_i) Q(z_i)\right)$ It is easy to verify that (a) and (b) above imply

$$(1 - 1/n)\|q\| \leq \|Pq\| \leq (1 + 1/n)\|q\| \quad \text{all } q \text{ in } *G$$

and $P(q) = *\Phi(q)$ for all q in $*(G \cap E)$.

The following result with condition (ii) deleted is an immediate consequence of Proposition 3.12 and the Principle of Local Reflexivity (Proposition 3.7). The condition on the dual space may be obtained by paralleling the argument above with Φ the identity transformation, with the sets $A(n,G)$ replaced by $A(n,G,S)$ where S is a finite subset of E' and with the additional restriction that

$$|\langle *y,q\rangle - \langle P(q),*y\rangle| \leq \frac{1}{n}\|q\|$$

for all q in $*G$ and y in S.

<u>Proposition 3.13.</u> Assume $*M$ is κ-saturated, $E \subseteq \hat{X}$ and card(E")

$< \kappa$. There is an isometry T of E'' into $\hat{X}$ which satisfies (i) $Tx = x$ for all $x \in E$ and (ii) $\langle y,x\rangle = \langle Tx,y\rangle$ for all $x \in E''$ and $y \in E'$.

Clearly, in the result above, E'' may be replaced by any even dual space. As an example of the application of Proposition 3.13 to the study of the structure of nonstandard hulls, we give the following.

Corollary 3.14. If $^*M$ is κ-saturated where κ is greater than the cardinality of the continuum, then c_0 is not isometric to a complemented subspace of any nonstandard hull $\hat{X}$.

Proof. Suppose Φ were an isometry of c_0 onto a complemented subspace S of $\hat{X}$ and P is the projection of $\hat{X}$ onto S. Then by Proposition 3.13 there is an extension Ψ of Φ which embeds ℓ_∞ isometrically into $\hat{X}$. It follows that $\Phi^{-1} P \Psi$ is a bounded projection of ℓ_∞ onto c_0, which is impossible [LT 2].

An illustration of the use of reflecting subspaces, which is proved using Proposition 3.11, is the following: (a uniform Hahn-Banach extension from E to F.)

Corollary 3.15. If F is a standard Banach space and E is a reflecting subspace of F, then there is a linear isometry T from E' into F' which has the property that for each $\phi \in E'$, $T(\phi) \in F'$ is an extension of ϕ to a linear functional defined on all of F.

Proof. (This can be proved by a weak *-compactness argument.) Choose M to contain E and F and choose $^*M$ to be κ-saturated, where $\kappa >$ cardinality of F. By Proposition 3.12, letting Φ be the inclusion of E into $\hat{E}$, there is an extension Ψ of Φ which is a linear isometry from F onto a subspace of $\hat{E}$. Since $E' \subseteq \widehat{E'} \subseteq (\hat{E})'$, we may define $T: E' \to F'$ by letting $\langle T(\phi),x\rangle = \langle \phi,\Psi(x)\rangle$ for each $\phi \in E'$ and $x \in F$. It is now easy to check that T has the desired properties.

Unless otherwise noted the results in this Section are due to the authors. (See [HMR 2], [HMR 3].) J. Stern [STE 2] obtained Theorem 3.2 independently in the context of ultrapowers. Also Proposition 3.12 and 3.8 show that E is a reflecting subspace of F if and only if F is a u-extension of E in the sense introduced by Stern [STE 4]. Stern independently proved results that are equivalent to Proposition 3.8 and 3.12 and to special cases of Proposition 3.13 and Corollary

3.14. W. A. J. Luxemburg gave an application of Helley's Theorem which is part of Proposition 3.13. (See [LUX 2].)

Section 4. Banach Lattices.

We noted in Section 1 that the nonstandard hull $\hat{L}$ of a Banach lattice L receives a natural order structure making it again a Banach lattice and L a Riesz subspace (see below). More generally if W is an internal Banach lattice, i.e., $W \in {}^*F$ where F is a family of Banach lattices in M, then $\hat{W}$ is again a standard Banach lattice under the ordering $\pi(p) \leq \pi(q)$ if and only if there exists h in $\mu(W)$ such that $p \leq q + h$.

We review some elementary notation and terminology. (See [LXZ] or [LT 3].) A normed Riesz space is a vector lattice L which is a normed space and such that

(i) $u \leq v$ implies $u + w \leq v + w$ all u,v,w in L,

(ii) $u \leq v$ and $0 \leq \lambda \in \mathbb{R}$ implies $\lambda u \leq \lambda v$ all u,v in L,

(iii) $|x| \leq |y|$ implies $\|x\| \leq \|y\|$ all x,y in L.

Thus a Banach lattice is a norm-complete normed Riesz space.

If L is a normed Riesz space, a subspace A is called a Riesz subspace if it is closed under the lattice operations. For a sequence $\{u_n\}$ (a net $\{u_\lambda\}$) in L we write $u_n \downarrow$ $(u_\lambda \downarrow)$ if $u_{n+1} \leq u_n$ for all n (if for every λ_1, λ_2 there exists λ_3 such that $u_{\lambda_3} \leq \inf(u_{\lambda_1}, u_{\lambda_2})$). If in addition $\inf_n u_n = 0$ $(\inf_\lambda u_\lambda = 0)$ we write $u_n \downarrow 0$ $(u_\lambda \downarrow 0)$.

The first question we investigate is the extent to which the finite dimensional Riesz subspaces of a Banach lattice play a role analogous to that played by the finite dimensional subspaces of a general Banach space. Recall that for every Banach space X in M there exists a *-finite dimensional subspace S of ${}^*X$ such that $X \subseteq \hat{S}$ (Lemma 1.4.).

Definition 4.1. Let F be the family of finite-dimensional Riesz subspaces of a Banach lattice L. Then if $W \in {}^*F$, we call $\hat{W}$ a hyperfinite dimensional Riesz subspace of $\hat{L}$.

Note that a hyperfinite dimensional Riesz subspace of $\hat{L}$ is both a hyperfinite dimensional subspace and a Riesz subspace of $\hat{L}$. It is natural to expect that for every Banach lattice there exists a hyperfinite dimensional Riesz subspace $\hat{W}$ of $\hat{L}$ such that $L \subseteq \hat{W} \subseteq \hat{L}$. In general this is false. One must assume L contains enough finite dimensional Riesz subspaces.

Definition 4.2. A normed Riesz space is said to be rich in finite dimensional Riesz subspaces if for every finite subset $\{x_1, x_2, \ldots, x_n\}$ of L and every $\delta > 0$ there exists a finite dimensional Riesz subspace F of L and $f_1, \ldots, f_n$ in F such that $\| x_i - f_i \| < \delta$ for $i = 1,2,\ldots,n$.

Proposition 4.3. Let L be a Banach lattice. Then L is contained in a hyperfinite dimensional Riesz subspace of $\hat{L}$ if and only if L is rich in finite dimensional Riesz subspaces.

Before we give a proof of Theorem 4.3, we show that $C[0,1]$ is not rich in finite dimensional Riesz subspaces. Consider the two element set consisting of the function u with constant value one and the identity function e, i.e., $e(x) = x$ for all x in $[0,1]$. Suppose F is a finite dimensional Riesz subspace of $C[0,1]$ containing f_1, f_2 so that $\| f_1 - a \| < 1/3$ and $\| f_2 - e \| < 1/3$. Then F must be at least two-dimensional. Otherwise $f_2 = \lambda f_1$ for some scalar λ and $|f_2(0)| < 1/3$ and $|f_1(0)| > 2/3$ imply $|\lambda| < 1/2$. But $f_2(1) > 2/3$ and $f(1) < 4/3$ which produces a contradiction. Now F has a basis consisting of functions with pairwise disjoint support, so for some x_o in $[0,1]$, $f(x_o) = 0$ for all f in F. Then $\| u - f_1 \| \geq | 1 - f(x_o) | = 1$ which is a contradiction.

Proof of Proposition 4.3. Assume L is rich in finite dimensional Riesz subspaces. Let G be a *-finite set containing $\{{}^*f : f \in L\}$ and let η be a positive infinitesimal. By the Transfer Principle there exists a *-finite dimensional Riesz subspace S of $^*L$ such that for every g in G there exists p in S with $\| g - p \| < \eta$. It follows immediately that $L \subseteq \hat{S}$.

Assume now that for some *-finite dimensional Riesz subspace S of $^*L$, we have $L \subseteq \hat{S}$. Let $\{x_1, x_2, \ldots, x_n\}$ be a finite subset of L and δ a positive real number. There exist $p_1, p_2, \ldots, p_n$ in S such that $\| {}^*x_i - p_i \| < \delta$ for $i = 1,2,\ldots,n$. It follows from the Transfer Principle, used in the other direction, that there exist a finite dimensional Riesz subspace F of L and $f_1, f_2, \ldots, f_n$ in F such that $\| x_i - f_i \| < \delta$ for $i = 1,2,\ldots,n$.

We recall various order-completeness conditions on a normed Riesz space L.

Definition 4.4. (i) L is said to have the principal projection property if for every $0 \leq u,v \in L$ there exists $P_u(v) = \sup_n \inf(nu,v)$.

(ii) L is said to be Dedekind complete (Dedekind σ-complete) if every nonempty (countable) subset of L which is bounded above has a supremum in L.

(iii) L is said to have an order-continuous norm (σ-order continuous norm) if $u_\lambda \downarrow 0$ $(u_n \downarrow 0)$ implies $\inf\|u_\lambda\| = 0$ $(\inf\|u_n\| = 0)$.

For a Banach lattice these properties are related as follows: order-continuous norm $\Rightarrow$ Dedekind complete $\Rightarrow$ Dedekind σ-complete $\Rightarrow$ principal projection property [LXZ].

It is a simple consequence of the Freudenthal Spectral Theorem [LXZ] that a normed Riesz space with the principal projection property is rich in finite dimensional Riesz subspaces. However, this condition is not necessary. Clearly if a normed Riesz space is rich in finite dimensional Riesz subspaces, so is its completion. Luxemburg [LUX 3] gives an example of a normed Riesz space L with the principal projection property but such that the norm-completion of L fails to have this property. It would be interesting to obtain a characterization of those Banach lattices which are rich in finite dimensional Riesz subspaces.

Definition 4.5. Let K and L be Banach lattices and let λ be a standard real number with $\lambda \geq 1$. Then

(i) L is said to be Riesz λ-embeddable in L if there exists an order isomorphism T of L into K such that $\|x\| \leq \|Tx\| \leq \lambda\|x\|$ for all x in L. In particular if $\lambda = 1$, we say that L is Riesz isometrically embeddable in K.

(ii) L is said to be Riesz finitely λ-representable in K if for each finite dimensional Riesz subspace F of L and each $\delta > 0$, the space F is Riesz $(\lambda + \delta)$-embeddable in L.

The proof of the following result is a simple modification of the proof of Proposition 3.2 and is omitted.

Proposition 4.6. Let L be a separable Banach lattice which is rich in finite dimensional Riesz subspaces and let $\hat{W}$ be the nonstandard hull of an internal Banach lattice. Then L is Riesz finitely λ-representable in $\hat{W}$ if and only if L is Riesz λ-embeddable in $\hat{W}$.

Again if we assume κ^+-saturation, we may replace the assumption of separability on L with the assumption that L have density character less than or equal to κ.

In general nonstandard hulls of internal Banach lattices have nice order properties.

Proposition 4.7. Let W be an internal Banach lattice. Then

(i) $\hat{W}$ has σ-order continuous norm, and

(ii) every increasing norm-bounded sequence in $\hat{W}$ is order-bounded.

Proof. (i) Suppose $u_n \downarrow 0$ in $\hat{W}$ but $\| u_n \| \geq \alpha > 0$ for all n. Pick $p_n \in W^+$ such that $p_n \downarrow \geq 0$ and $\pi(p_n) = u_n$ all n. Then $\| p_n \| > \alpha/2$ for all $n \in N$ and $A_n = \{p \in W: \| p \| > \alpha/2$ and $0 \leq p \leq p_n\}$ is a nonempty internal set for each finite n. Hence by $\aleph_1$-saturation there exists $p \in \bigcap_{n=1}^{\infty} A_n$. Then $0 < \pi(p) \leq \pi(p_n) = u_n$ which contradicts $u_n \downarrow 0$.

(ii) We may assume $u_n \uparrow$ in $\hat{W}$ and $\| u_n \| \leq 1$ for all finite n. Pick $0 \leq p_n \uparrow$ in W such that $\pi(p_n) = u_n$ for each n in N. Again by $\aleph_1$-saturation there exists $p \in W$ such that $p_n \leq p$ for all finite n and $\| p \| \leq 2$. Thus $p \in \text{fin}(W)$ and $u_n \leq \pi(p)$ for each n.

The following example shows that the nonstandard hull of an internal Banach space need not have order continuous norm or even be Dedekind σ -complete.

Example 4.8. We show that the nonstandard hull of c_0 is not Dedekind σ-complete. Let u_n be the element of c_0 given by

$$u_n(k) = \begin{cases} 1 & k \leq n \\ 0 & k > n \end{cases}$$

for $n = 1,2,\ldots$. Then $u_n \uparrow$ and it is clearly bounded above in $\hat{c}_0$ by say $\pi(p)$ where for some positive infinite integer θ

$$p(k) = \begin{cases} 1 & k \leq \theta \\ 0 & k > 0 \end{cases} .$$

We assert that $\{u_n\}$ does not have a supremum in $\hat{c}_0$. Suppose $\pi(q) \geq u_n$ for every finite integer n. Then $q(k) > 2/3$ for every finite integer k and so for some infinite integer ϕ. If q_1 is the element of $^*c_0$ defined by

$$q_1(k) = \begin{cases} q(k) & k \neq \phi \\ q(\phi) - 2/3 & k = \phi \end{cases}$$

then $u_n \leq \pi(q_1)$ for all $n \in N$ and $\pi(q_1) \lneq \pi(q)$.

Theorem 4.9. Let $\hat{W}$ be the nonstandard hull of an internal Banach lattice, then the following statements are equivalent:

(i) $\hat{W}$ has order-continuous norm,

(ii) $\hat{W}$ is Dedekind-complete,

(iii) $\hat{W}$ is Dedekind-σ-complete,

(iv) $\hat{W}$ has the principal projection property,

(v) c_0 is not Riesz isometrically embeddable in $\hat{W}$,

(vi) c_0 is not Riesz finitely 1-representable in $\hat{W}$.

Proof. As we noted above, in any Banach lattice (i) $\Rightarrow$ (ii) $\Rightarrow$ (iii) $\Rightarrow$ (iv) and the equivalence of (v) and (vi) follows from Proposition 4.6 since c_0 is rich in finite dimensional Riesz subspaces. The fact that (vi) and (i) are equivalent in any Banach lattice is a theorem of Meyer-Nieberg [MN].

Thus it remains to show that if c_0 is Riesz isometrically embeddable in $\hat{W}$, then $\hat{W}$ does not have the principal projection property. Let Φ be the order isometry of c_0 into $\hat{W}$. We may select $0 \leq p_n \in W$ for $n \in N$ so that $\inf(p_n, p_m) = 0$ for $n \neq m$ and $\pi(p_n) = \Phi(e_n)$ for each n in N where $\{e_n\}$ is the standard basis in c_0. Now by $\aleph_1$-saturation we may extend the sequence $\{p_n\}$ in W out to an infinite integer θ so that $0 \leq p_k$, $1 \leq k \leq \theta$, $\inf(p_j, p_k) = 0$ for $j \neq k$, $1 \leq j,k \leq \theta$ and if $p = \sum_{k=1}^{\theta} p_k$ then $\|p\| \approx 1$.

Let e be a positive weak unit in c_0, say $e(k) = 1/k$ for all c in N. We show that the set $P = \{\inf(n\,\Phi(e), \pi(p)) : n \in N\}$ does not have a supremum and so $\hat{W}$ does not have the principal projection property. Suppose to the contrary $\pi(q) = \sup P$ where $0 \leq q \in W$. Then $\pi(\inf(q, p_n)) = \inf(\pi(q), e_n) = e_n$ for each n in N, so $\|\inf(q, p_n)\| > 2/3$ for each finite n. It follows that for some infinite integer $\phi \leq \theta$, we have $\|\inf(q, p_\phi)\| > 2/3$. Now $\inf(n\,\Phi(e), \pi(p)) \leq \pi(q - \inf(q, p_\phi))$ for all finite n and $\pi(q - \inf(q, p_\phi)) = \pi(q) - \pi(\inf(q, p_\phi)) \lneq \pi(q)$ which is a contradiction.

<u>Definition 4.10.</u> Let K be a Banach lattice and let L be a closed Riesz subspace. Then L is said to be a <u>reflecting sublattice</u> of K if for every finite dimensional Riesz subspace F of K and every $\varepsilon > 0$ there is a Riesz isomorphism T of F into L such that

(i) $(1 - \varepsilon)\|x\| \leq \|Tx\| \leq (1 + \varepsilon)\|x\|$ all $x \in F$ and

(ii) $\|Tx - x\| \leq \varepsilon\|x\|$ all x in $F \cap L$.

Every Banach lattice L is a reflecting sublattice of its nonstandard hull. Indeed the following order analog to Proposition 3.8 can be established by a straightforward adaption of the proof.

<u>Proposition 4.11.</u> Let L be a Banach lattice, let F be a finite dimensional Riesz subspace of $\hat{L}$, and let $\{y_1, y_2, \ldots, y_m\}$ be a finite subset of L'. Then for every $\varepsilon > 0$ there exists a Riesz isomorphism T of F into L such that

(i) $(1 - \varepsilon)\|x\| \leq \|Tx\| \leq (1 + \varepsilon)\|x\|$ all x in F,

(ii) $\|Tx - x\| \leq \varepsilon\|x\|$ for all x in $L \cap F$, and

(iii) $|\langle x, y_i\rangle - \langle Tx, y_i\rangle| \leq \varepsilon\|x\|$ for all x in F and $i = 1, 2, \ldots, m$.

By another result of Meyer-Nieberg [MN] a Banach lattice is reflexive if and only if for all $\lambda \geq 1$ neither ℓ_1 nor c_0 is Riesz λ-embeddable in L. It is easy to verify that if c_0 (resp. ℓ_1) is Riesz λ-embeddable in L for some $\lambda \geq 1$ then c_0 (resp. ℓ_1) is Riesz finitely 1-representable in L. Combining these remarks, Propositions 4.6 and 4.10 and Corollary 3.9 we have the following characterizations of super-reflexive Banach lattices:

<u>Proposition 4.12.</u> If L is a Banach lattice, the following statements are equivalent:

(i) L is super-reflexive.

(ii) $\hat{L}$ is reflexive.

(iii) Neither ℓ_1 nor c_0 is Riesz finitely 1-representable in L.

(iv) Neither ℓ_1 nor c_0 is Riesz isometrically embeddable in $\hat{L}$.

It is clear that if K is a Banach lattice which is rich in finite dimensional Riesz subspaces, then every reflecting sublattice of K is again rich in finite dimensional Riesz subspaces. Since for any Banach lattice L, the second dual L" is Dedekind complete, L" is always rich in finite-dimensional Riesz subspaces. Thus again $C[0,1]$ provides an example, this time showing that a Banach lattice need not be a reflecting sublattice of its second dual.

The following order-theoretic version of the Principle of Local Reflexivity is due to Conroy and the second author [CNM]. (See also [CON].) A nice proof is given by Bernau in [BER].

Theorem 4.13. Let L be a Banach lattice, let F be a finite-dimensional Riesz subspace of L'', and let $\{y_1,y_2,\dots,y_n\}$ be a finite set of elements of L'. Then for every $\varepsilon > 0$ there exists a Riesz isomorphism T of F into L such that

(i) $(1-\varepsilon)\|x\| \le \|Tx\| \le (1+\varepsilon)\|x\|$ for all x in F,

(ii) $|\langle Tx,y_k\rangle - \langle x,y_k\rangle| \le \varepsilon\|x\|$ for all x in F and $k = 1,2,\dots,n$.

If L has order-continuous norm, then for every $\varepsilon > 0$ there exists a Riesz isomorphism T satisfying (i) and (ii) and

(iii) $\|Tx - x\| \le \varepsilon\|x\|$ for all x in $L \cap F$.

In particular every Banach lattice with order-continuous norm is a reflecting sublattice of its second dual.

The proof of the following proposition is analogous to the proof of Proposition 3.12 and is omitted.

Proposition 4.14. Let $\hat{W}$ be the nonstandard hull of an internal Banach lattice, let L be a Banach lattice and let Φ be a Riesz isometry of L into $\hat{W}$. If $^*M$ is κ-saturated and if L is a reflecting sublattice of a Banach lattice K where K is rich in finite dimensional Riesz subspaces and $\operatorname{card} K < \kappa$, then there is an extension Ψ of Φ to a Riesz isometry of K into $\hat{W}$.

Combining Proposition 4.13 and 4.14 we have:

Corollary 4.15. If L is a Banach lattice with order-continuous norm, then there is a Riesz isometry of L'' into $\hat{L}$ which leaves L invariant.

If L is a Banach lattice without an order-continuous norm, one can show that there is a Riesz isometry of L'' into $\hat{L}$ but it need not leave L invariant. In both cases one can also obtain the Riesz isometry T of L'' into $\hat{L}$ so that

$$\langle x,y\rangle = \langle Tx,y\rangle \quad \text{for all } x \text{ in } L'' \text{ and } y \text{ in } L'.$$

We now consider certain special classes of Banach lattices. Recall the following definitions:

Definition 4.16. Let L be a Banach lattice.

(i) For $1 \leq p < \infty$, L is said to be an abstract L^p-space if $0 \leq u,v \in L$ with $\inf(u,v) = 0$ implies $\|u + v\|^p = \|u\|^p + \|v\|^p$.

(ii) L is said to be an abstract M-space if $0 \leq u,v \in L$ with $\inf(u,v) = 0$ implies $\|u + v\| = \max(\|u\|, \|v\|)$.

If L is an abstract L^p-space, then for some measure space (X,Σ,ν), L is Riesz isometric to $L^p(X,\Sigma,\nu)$. If L is an abstract M-space, then L is Riesz isometric to a closed Riesz subspace of a space $C(X)$ where X is a compact Hausdorff space.

Proposition 4.17. (i) If W is an internal abstract L^p-space where p is a finite element of ${}^*\mathbb{R}$, then $\hat{W}$ is an abstract L^r-space where $r = st(p)$. In particular the nonstandard hull of a standard L^p-space $(1 \leq p < \infty$, p in $\mathbb{R})$ is again an abstract L^p-space.

(ii) If W is an internal abstract L^p-space where p is infinite or W is an internal abstract M-space, then $\hat{W}$ is an abstract M-space. In particular the nonstandard hull of a standard abstract M-space is again an abstract M-space.

Proof. (i) Let W be a internal abstract L^p-space where p is a finite element of ${}^*\mathbb{R}$ and set $r = st\, p$. Assume $0 \leq u,v \in \hat{W}$ and $\inf(u,v) = 0$. Then there exist $0 \leq p_1,p_2$ in $fin(W)$ such that $\inf(p_1,p_2) = 0$, $\pi(p_1) = u$ and $\pi(p_2) = v$. Hence

$$\|u + v\|^r = [st(\|p_1 + p_2\|)]^r = st(\|p_1 + p_2\|^p)$$

$$= st(\|p_1\|^p + \|p_2\|^p) = \|u\|^r + \|v\|^r .$$

(ii) If W is an internal abstract L^p-space where p is infinite and $0 \leq u,v \in \hat{W}$ with $\inf(u,v) = 0$, then again there exist p_1, p_2 in $fin(W)$ such that $\inf(p_1,p_2) = 0$, $\pi(p_1) = u$ and $\pi(p_2) = v$. Then

$$\|u + v\| = st(\|p_1 + p_2\|)$$

$$= st[(\|p_1\|^p + \|p_2\|^p)^{1/p}] = \max(st\|p_1\|, st\|p_2\|)$$

$$= \max(\|u\|, \|v\|).$$

The rest of the proof is trivial.

Included in Proposition 4.17 is the fact (proved by Dacunha-Castelle and Krivine [DCK 1]) that if p is a fixed number $1 \leq p < \infty$, then a Banach space ultraproduct of L_p-spaces is again an L_p-space.

The nonstandard hulls of standard L_p-spaces are described in detail in Section 6 and the nonstandard hulls of internal spaces $\ell_p(n)$ with n and p infinite are considered in Section 9. We close this section by showing that the nonstandard hull of an internal $C(K)$ space is (Riesz isometric to) a space of continuous functions on a compact set. In particular, the nonstandard hull of a standard $C(K)$ space is again a $C(K)$ space.

Let X be an internal compact Hausdorff space. The internal space $C(X)$ consists of internal continuous (Q-continuous in Robinson's terminology) functions mapping X into ${}^*\mathbb{R}$. An element f in $C(X)$ is norm-finite if and only if the range of f is contained in $\mathrm{fin}({}^*\mathbb{R})$. For each such f we define a pseudometric d_f on X by $d_f(x,y) = \mathrm{st}|f(x) - f(y)|$ and designate by $\mathcal{U}$ the uniformity on X generated by all such pseudometrics. Let $\hat{X}$ be the completion of X with respect to $\mathcal{U}$.

<u>Proposition 4.18.</u> The space $\hat{X}$ is a compact Hausdorff space and $C(X)^{\wedge}$ is Riesz isometric to $C(\hat{X})$.

<u>Proof.</u> If $x,y \in X$ with $x \neq y$ then by the Transfer Principle there exists $f \in \mathrm{fin}(C(X))$ such that $f(x) = 1$ and $f(y) = 0$. Thus $\mathcal{U}$ is a Hausdorff uniformity on X and $\hat{X}$ is also Hausdorff. In order to show that $\hat{X}$ is compact, it is sufficient to show that X is totally bounded. But let $f \in \mathrm{fin}(C(X))$ and say a,b are standard real numbers such that $a \leq f(x) \leq b$ for all x. If ε is a positive standard real, let $\{a = t_0 < t_1 < \ldots < t_n = b\}$ be a finite partition of $[a,b]$ such that $t_i - t_{i-1} < \varepsilon/3$ for $i = 1,2,\ldots,n$. If $\{x\colon t_{i-1} \leq f(x) < t_i\}$ is not empty, pick x_i in this set and let $X_i = \{x \in X\colon |f(x) - f(x_i)| < \varepsilon/2\}$. If the initial set is empty, let $X_i = \emptyset$, $i = 1,2,\ldots,n$. Then $\bigcup_{i=1}^{n} X_i$ is a decomposition of X into sets with d_f-diameter less than ε. It follows that X is totally bounded.

For each $p \in \mathrm{fin}(C(X))$, $\pi(p)$ in $C(X)^{\wedge}$ can be identified with the function which assigns $\mathrm{st}\, p(x)$ to each x in X. Since each $\pi(p)$ is clearly uniformly continuous, it extends uniquely to $\hat{X}$, indeed we have a Riesz isometry of $C(X)^\wedge$ onto a closed Riesz subspace of $C(\hat{X})$. Moreover, $C(X)^\wedge$ is naturally an algebra of functions and the mapping into $C(\hat{X})$ takes it onto a closed subalgebra containing the constant functions and separating points of $\hat{X}$. Thus by the Stone-Wierstrass Theorem we have that $C(X)^\wedge$ and $C(\hat{X})$ are Riesz isometric.

Except as otherwise indicated, the results in this section through 4.11 are due to Cozart and the second author [CZM]. Proposition 4.17 is essentially contained in [HMR 2] while Proposition 4.18 appears in [HEN 3].

Section 5. Some Model Theory.

A very useful tool for analyzing the structure of nonstandard hulls has been to regard Banach spaces as structures for a certain first-order language and to study their model-theoretic properties. This was done first in [HEN 2] to obtain a characterization of those pairs of standard spaces E,F whose nonstandard hulls $\hat{E},\hat{F}$ are linearly isometric. A model-theoretic approach was used in [HEN 1], where the isomorphism properties (a class of strong saturation properties) were introduced. (These will be discussed below.) In [HEN 3] this model-theoretic approach to nonstandard hulls was developed quite fully. (This approach has in fact led to a quite general and broadly based way to apply all model-theoretic constructions to Banach space theory, not just the ultraproduct and saturated model methods used in nonstandard analysis. See the forthcoming papers [HEN 5], [HH] for this development.)

Let L be the first-order language whose nonlogical symbols are a binary function symbol $+$, two unary predicate symbols P and Q and for each rational number r a unary function symbol f_r. Each real Banach space is regarded as an L-structure by taking $+_E$ to be addition on E, by setting $P_E = \{x: \|x\| \leq 1\}$ and $Q_E = \{x: \|x\| \geq 1\}$ and by taking $(f_r)_E$ to be the operation of scalar multiplication by r for each rational number r. A term t in L may be considered as an expression of the form $r_1x_1 + r_2x_2 + \ldots + r_nx_n$ where $r_1, r_2,\ldots,r_n$ are rational and $x_1,x_2,\ldots,x_n$ are variables. Then the interpretations in a Banach space E of the atomic formulas $P(t)$ and $Q(t)$ with $a_1,a_2,\ldots,a_n$ assigned to $x_1,x_2,\ldots,x_n$ are $\|r_1a_1 + r_2a_2 + \ldots + r_na_n\| \leq 1$ and $\|r_1a_1 + r_2a_2 + \ldots + r_na_n\| \geq 1$ respectively. A positive bounded formula in L is one which can be built up from atomic formulas using conjunction, disjunction and the bounded quantifiers $(\exists x)(Px \wedge \ldots)$ and $(\forall x)(Px \rightarrow \ldots)$. We will use the notation $\exists_B x \ldots$ for $\exists x(Px \wedge \ldots)$ and $\forall_B x \ldots$ for $\forall x(Px \rightarrow \ldots)$.

Consider for example the positive bounded sentence τ given by:

$$\exists_B x\, \exists_B y (Q(x) \wedge Q(y) \wedge P(x+y)$$
$$\wedge P(x-y) \wedge Q(x+y) \wedge Q(x-y)).$$

In any Banach space this asserts the existence of elements x and y such that $\|x\| = \|y\| = \|x+y\| = \|x-y\| = 1$. That is, x and y span a subspace that is linearly isometric to $\ell_1(2)$.

Now let σ be a positive bounded formula of L and let m be a positive integer. We obtain σ_m^+ from σ as follows: from atomic formulas (i) replace $t = s$ by $P(m \cdot (t-s))$; (ii) replace $P(t)$ by $P((1-(1/m)) \cdot t)$; (iii) replace $Q(t)$ by $Q((1+(1/m)) \cdot t)$. For more complex formulas let $(\sigma \wedge \tau)_m^+ = \sigma_m^+ \wedge \tau_m^+$, $(\sigma \vee \tau)_m^+ = \sigma_m^+ \vee \tau_m^+$, $((\exists_B x)\sigma)_m^+ = (\exists_B x)\sigma_m^+$, $((\forall_B x)\sigma)_m^+ = (\forall_B x)\sigma_m^+$ and proceed inductively. For example, if τ is the positive bounded sentence above, then τ_m^+ asserts the existence of x,y such that $1-\delta \le \|x\|, \|y\| \le 1$ and $1-\delta \le \|x \pm y\| \le 1+\varepsilon$ where $\delta = 1/(m+1)$ and $\varepsilon = 1/(m-1)$. As $m \to \infty$ this comes closer and closer to the condition expressed by τ; however, the x and y asserted to exist by the sentences τ_m^+ may depend on m.

Given a positive bounded sentence σ in L and a Banach space X, we say that σ is <u>approximately satisfied</u> in X if σ_m^+ holds in X for every positive integer m in N; in this case we write $X \models_A \sigma$. Two Banach spaces X and Y are <u>approximately equivalent</u> if for each positive bounded sentence σ in L, we have that σ is approximately satisfied in X if and only if σ is approximately satisfied in Y; in this case we write $X \equiv_A Y$. The most important instance of this relation is that $E \equiv_A \hat{E}$ holds for any standard Banach space E and any nonstandard hull $\hat{E}$ of E. (See Proposition 5.4.) If M possesses the $\aleph_0$-isomorphism property, the following results hold [HEN 3]:

<u>Proposition 5.1.</u> For standard Banach spaces E and F the following are equivalent:

(i) $E \equiv_A F$,

(ii) $\hat{E} \equiv_A \hat{F}$,

(iii) $\hat{E}$ is isometric to $\hat{F}$.

Proposition 5.2. For internal Banach spaces W and V in $^*M$ the following are equivalent:

(i) $\hat{W} \equiv_A \hat{V}$,

(ii) $\hat{W}$ is isometric to $\hat{V}$.

Another very different standard condition, which is equivalent to (i), (ii), (iii) of Proposition 5.1, is discussed below in Section 8. (See Theorem 8.1.)

Another useful model theoretic concept applies to subspaces E of Banach spaces F. We call E an elementary subspace of F and write $E \prec_A F$ if $E \subseteq F$ and for each positive bounded formula $\sigma(x_1,\ldots,x_n)$ and each $a_1,\ldots,a_n \in E$,

$$E \models_A \sigma(a_1,\ldots,a_n) \Leftrightarrow F \models_A \sigma(a_1,\ldots,a_n)$$

Evidently $E \prec_A F$ implies $E \equiv_A F$.

Proposition 5.3. For standard Banach spaces $E \subseteq F$, the following are equivalent:

(i) $E \prec_A F$.

(ii) There is a linear isometry Φ of $\hat{E}$ onto $\hat{F}$ such that $\Phi(x) = x$ for all $x \in E$.

This can be proved in essentially the same way as Proposition 5.1, except that the elements of E must be treated as part of the basic formal language ("constants"). The condition (ii) of Proposition 5.3 was studied in [HEN 1], which pre-dates the introduction of positive bounded formulas.

Note that Proposition 5.3 shows that whenever $E \subseteq F$ and $E \prec_A F$, then E is a reflecting subspace of F. (Use Proposition 3.8.)

The most important instance of the relation $\prec_A$ is given in the following (See [HEN 3, Corollary 1.10].):

Proposition 5.4. For any Banach space E and any nonstandard hull $\hat{E}$ of E,

$$E \prec_A \hat{E}$$

In particular $E \equiv_A \hat{E}$.

Remark. (About the $\aleph_0$-isomorphism Property) This property is simply a technical means for avoiding arguments that require nonstandard hulls of spaces which have themselves been constructed as nonstandard hulls. Strictly speaking, if E is a standard Banach space in M and $\hat{E}$ is its nonstandard hull, obtained from $*E$ in $*M$, then $\hat{E}$ is not in M. Therefore if we wish to apply a nonstandard analysis argument to the space $\hat{E}$, we must shift to a new M which contains $\hat{E}$ and then change $*M$ too. This is inconvenient and in Banach space theory it would occur rather often.

What happens if $*M$ satisfies the $\aleph_0$-isomorphism property is that any nonstandard hull $\hat{V}$ which is $\equiv_A$ to $\hat{E}$ will be linearly isometric to $\hat{E}$. It turns out in many arguments that this is a sufficient replacement for the "double nonstandard hull."

In most of the remainder of this paper we assume that $*M$ has the $\aleph_0$-isomorphism property. Usually we make explicit mention of the fact it is being used in the proof.

A number of useful model-theoretic tools have their counterparts in the Banach space setting.

Theorem 5.5. (Compactness Theorem) Let Σ be a set of positive bounded sentences which is finitely satisfiable in the sense that for each finite set $F \subseteq \Sigma$ and each integer m, there is a standard Banach space $E = E(F,m)$ such that σ_m^+ is true in E for each $\sigma \in F$.

Then there is a standard space E such that for every $\sigma \in \Sigma$ and every integer m, σ_m^+ is true in E. (In brief, $E \models_A \Sigma$.)

Theorem 5.6. (Downward Löwenheim-Skolem Theorem) Let E be a standard Banach space and κ an infinite cardinal number. For each set $S \subseteq E$ of cardinality $\leq \kappa$ there is a closed subspace F of E such that $S \subseteq F$, F has density character $\leq \kappa$ and $F \prec_A E$. (So, in particular, $F \equiv_A E$.)

For proofs of these results see [HEN 3, Section 1].

The following observation, shows how these ideas can interact.

Corollary 5.7. Assume that $*M$ has the $\aleph_0$-isomorphism property. For any internal Banach space V there exists a separable standard Banach space E such that $\hat{V}$ and $\hat{E}$ are isometric.

Proof. By Theorem 5.6 there is a separable space E such that $E \equiv_A \hat{V}$. Hence $\hat{E} \equiv_A \hat{V}$ by Proposition 5.4 so that $\hat{E}$ is isometric to $\hat{V}$ by Proposition 5.2.

Note also that the subspace F in Theorem 5.6 must be, in particular, a reflecting subspace of E. (See the comments after Proposition 5.3.)

We remark that downward Löwenheim-Skolem theorems stronger than

Theorem 5.6 are also useful. For example, in [HEN 1] such a result was used to obtain many projections on weakly compactly generated spaces, in greater generality than those constructed by Amir and Lindenstrauss [AML]. (See also the proofs of Theorems 6.7 and 6.8 and of Lemma 11.1 below.)

Many classes $\mathcal{C}$ of Banach spaces turn out to be <u>axiomatizable</u> in the sense that there exists a set Σ of positive bounded sentences with the property that for any Banach space E,

$$E \in \mathcal{C} \iff E \models_A \Sigma .$$

(Recall that $E \models_A \Sigma$ means that $E \models_A \sigma$ for all $\sigma \in \Sigma$.) We will give many examples of axiomatizable classes of classical Banach spaces in Section 6. The main tool for easily proving that a given class $\mathcal{C}$ is axiomatizable is the following result: (See [HEN 3, Theorem 1.20].)

<u>Theorem 5.8.</u> Let $\mathcal{C}$ be a class of Banach spaces, satisfying these conditions:

(a) if $E \in \mathcal{C}$ and F is isometric to E, then $F \in \mathcal{C}$;

(b) if E is a standard Banach space and some nonstandard hull of E is in $\mathcal{C}$, then E is in $\mathcal{C}$;

(c) if $\mathcal{B} \subseteq \mathcal{C}$ is a set in some superstructure M, $^*\mathcal{B}$ is the $*$-transform of $\mathcal{B}$ in a nonstandard extension $^*M$ and V is an internal Banach space in $^*\mathcal{B}$, then $\hat{V} \in \mathcal{C}$. Under these hypotheses, $\mathcal{C}$ is an axiomatizable class of Banach spaces.

<u>Remark.</u> Condition (c) in Theorem 5.8 is equivalent to the assertion that $\mathcal{C}$ is closed under the formation of Banach space ultraproducts.

Finally we consider the behavior of the equivalence relation $\equiv_A$ under passage to dual spaces. For super-reflexive spaces, the situation is very nice.

<u>Theorem 5.9.</u> If E, F are super-reflexive Banach spaces, then

$$E \equiv_A F \iff E' \equiv_A F' .$$

<u>Proof.</u> Since $E = (E')'$ and the same for F, it suffices to prove $\Rightarrow$. If $E \equiv_A F$, then $\hat{E}, \hat{F}$ are isometric (if we choose $^*M$ correctly, as we may). Hence $(\hat{E})'$ and $(\hat{F})'$ are also isometric. But by Proposition 3.11 we have that $\widehat{(E')}$ and $\widehat{(F')}$ are isometric and hence that $E' \equiv_A F'$ (using Proposition 5.1 again).

For arbitrary E and F, the equivalence in Theorem 5.9 does not hold. For example, let E be ℓ_1 and let F be $L^1(\mu)$, where μ

is a measure with infinitely many atoms but also a nontrivial nonatomic part. Then $E \equiv_A F$ by Theorem 6.1. On the other hand, E' is ℓ_∞ while F' is of the form $C(K)$, where K is a totally disconnected compact space which does not have a dense set of isolated points. By the remarks after Theorem 6.2 we see $E' \not\equiv_A F'$.

We can conclude from $E \equiv_A F$ that E' and F' are finitely represented in each other. This follows from the local duality theorem due to Kürsten and Stern that is discussed in Section 7. (See Corollary 7.2.) Perhaps a little more than this is true in general, but it is not clear how much.

Section 6. Classical Banach Spaces.

In this Section we will briefly summarize what is known about the nonstandard hulls of the classical Banach spaces (i.e., the spaces that are closely related to L_p-spaces and $C(K)$ spaces.) We will omit most proofs. This topic was treated in some detail in the survey paper by S. Heinrich [HEI 4], although we discuss some recent results that were not mentioned there.

In Section 4 it is shown that the classes of L_p-spaces $(p < \infty)$ and of $C(K)$-spaces are each closed under the formation of nonstandard hulls. (See Propositions 4.17 and 4.18.) It is also true that if E is a standard Banach space whose nonstandard hull $\hat{E}$ is isometric to an L_p-space $(p < \infty)$ or to a $C(K)$-space, then E satisfies exactly the same condition. For the L_p-spaces this was proved in [HMR 2]. For $C(K)$-spaces this is a recent important result due to S. Heinrich [HEI 5].

Theorem 6.1. Fix $1 \leq p < \infty$ and let μ, ν be two measures whose L_p-spaces are infinite dimensional. Then $L_p(\mu) \equiv_A L_p(\nu)$ if and only if either μ, ν both have an infinite number of atoms or they have the same finite number of atoms.

For example, Theorem 6.1 implies that if μ is any measure with an infinite number of atoms and $1 \leq p < \infty$, then the nonstandard hull of $L_p(\mu)$ is linearly isometric to $\hat{\ell}_p$. (See Proposition 5.1.) Moreover, every Banach space whose nonstandard hull is isometric to $\hat{\ell}_p$ must be of this form.

The classification of $C(K)$ spaces under $\equiv_A$ is incomplete. However it is known in the case where K is a totally disconnected compact space. Given such a space, let $B(K)$ denote the Boolean algebra of all clopen subsets of K.

<u>Theorem 6.2.</u> Let K_1 be a totally disconnected compact space and let K_2 be any other compact space.

(a) If $C(K_1) \equiv_A C(K_2)$ then K_2 is also totally disconnected.

(b) If K_1, K_2 are both totally disconnected, then $C(K_1) \equiv_A C(K_2)$ if and only if the Boolean algebras $\mathcal{B}(K_1)$ and $\mathcal{B}(K_2)$ satisfy exactly the same sentences in the first-order language of Boolean algebras.

It follows from Theorem 6.2, for example, that a Banach space E has its nonstandard hull isometric to $\hat{\ell}_\infty$ if and only if E is isometric to a $C(K)$-space where K is totally disconnected and compact and K has a dense set of isolated points.

Theorems 6.1 and 6.2 were first proved in [HEN 3], where some further structural information about the nonstandard hulls of L_p- and $C(K)$-spaces can be found.

A straightforward way to express some results classifying Banach spaces under $\equiv_A$ is in terms of axiomatizable classes of spaces. We use Theorem 5.8 in proving such results (and also in interpreting their significance.)

<u>Theorem 6.3.</u> Each class of Banach spaces mentioned below is axiomatizable:

(1) (Fix $1 \leq p < \infty$) the L_p-spaces;

(2) (Fix $1 \leq p < \infty$) the L_p-spaces based on an atomless measure;

(3) the $C(K)$-spaces;

(4) the $C(K)$-spaces for which K is totally disconnected;

(5) the $C(K)$-spaces for which K is connected;

(6) (Fix $1 \leq p \leq \infty$ and $\lambda \geq 1$) the $\mathcal{L}_{p,\lambda+}$-spaces.

<u>Discussion of Proofs.</u> In each case we must verify the conditions (b) and (c) of Theorem 5.8. (Condition (a) is obvious.) For cases (1)-(4) this is covered by Theorems 6.1 and 6.2 and the discussion preceding them. Case (5) is covered by these results and [HEN 3, Corollary 3.4]. For case (6) the details may be found in [HMR 2]. A Banach space E is a $\mathcal{L}_{p,\lambda+}$-space if for each $\varepsilon > 0$ and each finite dimensional subspace $X \subseteq E$, there exists a finite dimensional $Y \subseteq E$ such that $X \subseteq Y$ and Y is $(\lambda + \varepsilon)$-isomorphic to $\ell_p(n)$, $n = \dim(Y)$.

For a class C of Banach spaces to be closed under ultraproducts (or under nonstandard hull formation as given in clause (c) of Theorem 5.8) is a kind of compactness property of C. Such a property can often be used to obtain uniform bounds. For example, part (6) of Theorem

6.3 yields that, in the definition of $\mathcal{L}_{p,\lambda+}$-spaces given above, one may require that there be a uniform bound on the dimension of Y as a function of $\dim(X)$ and ε, without changing the meaning of the concept.

The $\mathcal{L}_{\infty,1+}$-spaces are just the L_1-preduals. An interesting class of such spaces is the class of Gurarii spaces:

Definition 6.4. A Banach space E is a Gurarii space [GU] if for each finite dimensional $X \subseteq E$, each finite dimensional $Y \supseteq X$ and each $\varepsilon > 0$, there exists a $(1 + \varepsilon)$-isomorphism T from Y onto a subspace of E such that $T(x) = x$ for all $x \in X$.

Usually only the separable Gurarii spaces are considered. It is immediate from the definition that if E and F are separable Gurarii spaces then E, F are $(1 + \varepsilon)$-isomorphic for each $\varepsilon > 0$. (In fact any such separable Gurarii spaces must be isometric [LUS].)

Theorem 6.5. Let E, F be Banach spaces.

(a) If E is a Gurarii space and $F \equiv_A E$, then F is also a Gurarii space;

(b) if E, F are both Gurarii spaces, then $E \equiv_A F$.

Proof. (a) If $E \equiv_A F$ then $\hat{E}$ and $\hat{F}$ are isometric. Thus it suffices to show that for any E,

$$E \text{ is a Gurarii space} \iff \hat{E} \text{ is a Gurarii space.}$$

The implication $\Rightarrow$ is a simple transfer argument. The other implication is immediate from the fact that E is a reflecting subspace of $\hat{E}$. (Proposition 3.8.)

(b) Suppose E, F are Gurarii spaces. Use the Downward Löwenheim-Skolem Theorem (Theorem 5.6) to obtain separable spaces E_0 and F_0 so that $E_0 \equiv_A E$ and $F_0 \equiv_A F$. The result of Lusky [LUS] yields that E_0 and F_0 are isometric, since they are Gurarii spaces by (a). Hence $E \equiv_A F$. (Actually $E_0 \equiv_A F_0$ is implied immeidately from the fact that they are $(1 + \varepsilon)$-isomorphic for each $\varepsilon > 0$. Taking $\varepsilon > 0$ to be infinitesimal yields that $\hat{E}_0$ and $\hat{F}_0$ are isometric and hence $E_0 \equiv_A F_0$.)

It is not difficult to construct directly a set Σ of positive bounded sentences which axiomatizes the class of Gurarii spaces. Given finite dimensional spaces $X \subseteq Y$, one chooses a nice basis $a_1,\ldots,a_n$

for X and $a_1,\ldots,a_n,b_1,\ldots,b_m$ for Y; then given $\varepsilon > 0$ one constructs a positive bounded sentence

$$\forall_B x_1 \ldots \forall_B x_n \exists_B y_1 \ldots \exists_B y_m (A(\vec{x}) \to B(\vec{x},\vec{y}))$$

where $A(\vec{x})$ is a quantifier-free formula almost exactly expressing that the sequence $x_1,\ldots,x_n$ is $(1+\frac{\varepsilon}{2})$-equivalent to $a_1,\ldots,a_n$ and $B(\vec{x},\vec{y})$ is a quantifier-free formula almost exactly expressing that $x_1,\ldots,x_n,y_1,\ldots,y_m$ is $(1+\varepsilon)$-equivalent to $a_1,\ldots,a_n$, $b_1,\ldots,b_m$. Similar ideas can be used in obtaining explicit axioms for L_p-spaces or for $\mathcal{L}_{p,\lambda+}$-spaces. However, no explicitly given set Σ of positive bounded sentences has been given which axiomatizes the class of $C(K)$-spaces. Perhaps such a set could be found by a close analysis of the ideas used in [HEI 5] by S. Heinrich.

In the remainder of this Section we consider isomorphic (as opposed to isometric) equivalence of nonstandard hulls. First we give some results proved by J. Stern [STE 4] in a somewhat improved form based on [HEN 3].

<u>Theorem 6.6.</u> Assume $^*M$ has the $\aleph_0$-isomorphism property and let E and F be infinite dimensional Banach spaces.

(a) If E and F are $\mathcal{L}_p$-spaces and $1 < p < \infty$, then $\hat{E}$ and $\hat{F}$ are isomorphic.

(b) If E and F are isomorphic to complemented subspaces of $L_1(\mu)$-spaces, then $\hat{E}$ and $\hat{F}$ are isomorphic.

(c) If E and F are isomorphic to complemented subspaces of $C(K)$-spaces, then $\hat{E}$ and $\hat{F}$ are isomorphic.

<u>Proof.</u> (a) Let E be a $\mathcal{L}_p$-space, $1 < p < \infty$. Theorem 4.4 of [STE 4] yields that E has a nonstandard hull which is isomorphic to $L_p(\mu)$ for some measure μ. By a downward Löwenheim-Skolem argument, applied simltaneously to this nonstandard hull of E, to $L_p(\mu)$ and to the isomorphism between them, one obtains a separable $E_0 \equiv_A E$ and a separable $L_p(\nu)$ such that E_0 and $L_p(\nu)$ are isomorphic. (Here we are using the remark before Theorem 6.1, that any space $\equiv_A L_p(\mu)$ is isometric to some $L_p(\nu)$.) But then $\hat{E}_0$ and $\hat{E}$ are isometric, while $\hat{E}_0$ is isomorphic to $\hat{L}_p(\nu)$. The same construction applies to F. Thus $\hat{E}$ and $\hat{F}$ are each isomorphic to nonstandard hulls of separable L_p-spaces. It follows from [HEN 3, Corollary 2.3] that they are isomorphic to each other. Parts (b) and (c) are proved in a similar way, combining [STE 4, Theorem 4.5] with [HEN 3, Corollary 2.3 and Corollary 3.11].

Note that Theorem 6.6 implies that (a) if E is a $\mathcal{L}_p$-space and $1 < p < \infty$ then $\hat{E}$ is isomorphic to $\hat{\ell}_p$; (b) if E is a complemented subspace of some $L_1(\mu)$-space then $\hat{E}$ is isomorphic to $\hat{\ell}_1$, and (c) if E is a complemented subspace of some $C(K)$-space, then $\hat{E}$ is isomorphic to $\hat{\ell}_\infty$.

In the Banach lattice context we have the following result related to part (c) of Theorem 6.6.

Theorem 6.7. Assume $^*M$ has the $\aleph_0$-isomorphism property. If E is a Banach lattice which is an abstract M-space, then $\hat{E}$ is isomorphic to $\hat{\ell}_\infty$, as a Banach space.

Proof. By a lattice version of Theorem 5.6 (Downward Löwenhein-Skolem Theorem) there is a separable Banach lattice E_0 such that $E_0 \equiv_A E$ and E_0 is an abstract M-space. A result due to Y. Benyamini [BEN] implies that E_0 is isomorphic to a $C(K)$ space, as a Banach space. Now apply Theorem 6.6(c) to get that $\hat{E}_0$ is isomorphic to $\hat{\ell}_\infty$. By Proposition 5.1, $\hat{E}_0$ and $\hat{E}$ are isometric, completing the proof.

Finally, continuing this theme, we answer a question raised by J. Stern [STE 4, Problem 4.2], by showing that there is an $\mathcal{L}_\infty$ space E such that no nonstandard hull of E is isomorphic to a $C(K)$-space.

Theorem 6.8. If E is a Gurarii space and $\hat{E}$ is any nonstandard hull of E, then $\hat{E}$ is not isomorphic to a complemented subspace of any $C(K)$-space.

Proof. Suppose otherwise. By taking a nonstandard hull of $\hat{E}$ and using Theorem 6.6(c), we see that E has a nonstandard hull that is isomorphic to $C(K)$ for some K. We may assume it is the given $\hat{E}$ without loss of generality. By a downward Lowenheim-Skolen argument, we may obtain a separable $E_0 \prec_A \hat{E}$ and a separable $F_0 \prec C(K)$ such that E_0 and F_0 are isomorphic. By the result of S. Heinrich [HEI 5] mentioned before Theorem 6.1, F_0 also is a $C(K)$-space. (This can be arranged for directly in the Löwenheim-Skolen argument. See [HEN 3, Theorem 3.9].) But E_0 is a separable Gurarii space by Theorem 6.5(a). This contradicts a result of Benyamini and Lindenstrauss [BNL, Corollary 2].

Section 7. The Kürsten-Stern Local Duality Theorem, with Applications.

In this Section we will give essentially all of the details of a proof, due to Stefan Heinrich [HEI 1 and 4] of an important result first proved for ultraproducts of Banach spaces by J. Stern [STE 4] and K. D. Kürsten [KU]. Then we will give a few examples of how to use this result, in order to show how important it is.

Theorem 7.1. Let V be an internal Banach space and let V' be its internal dual space. Suppose $X \subseteq \hat{V}$ and $Y \subseteq (\hat{V})'$ are finite dimensional.

Then there is a linear isometry T from Y into $\widehat{V'}$ such that

a) $\langle x,y \rangle = \langle x,T(y) \rangle$ for all $x \in X$, $y \in Y$

b) $T(y) = y$ if $y \in Y \cap \widehat{V'}$.

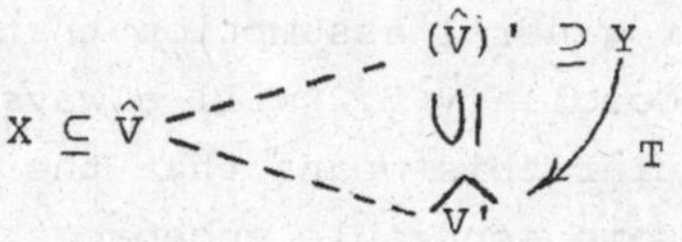

An immediate consequence of this result and the saturation principle in our nonstandard model is the following.

Corollary 7.2. Assume the nonstandard superstructure $^*M$ is κ-saturated ($\kappa \geq \aleph_1$). Let V be an internal Banach space and let V' be its internal dual space. Suppose $X \subseteq \hat{V}$ and $Y \subseteq (\hat{V})'$ are subspaces having topological density character $< \kappa$.

Then there exists a linear isometry T from Y into $\widehat{V'}$ such that

a) $\langle x,y \rangle = \langle x,Ty \rangle$ for all $x \in X$, $y \in Y$

b) $T(y) = y$ if $y \in Y \cap \widehat{V'}$.

Proof. Let $\langle v_i | i \in I \rangle$ be a subset of V whose image in $\hat{V}$ is dense in X and such that $\operatorname{card}(I) < \kappa$. Similarly let $\langle y_j | j \in J \rangle$ be a dense subset of Y such that $\operatorname{card}(J) < \kappa$ and such that for some subset J_0 of J the set $\langle y_j | j \in J_0 \rangle$ is dense in $Y \cap \widehat{V'}$. For each $j \in J_0$ select $u_j \in {}^*V'$ such that $\pi(u_j) = y_j$. We must find a set $\langle w_j | j \in J \rangle$ contained in V' satisfying these conditions:

(1) For each $i \in I$, $j \in J$ and each standard integer $n \in N$

$$|\langle v_i, w_j \rangle - \langle \pi(v_i), y_j \rangle| < 1/n.$$

(2) For every finite set $\{n_1,\ldots,n_k\} \subseteq J$, every finite sequence $r_1,\ldots,r_k$ of rational numbers and every standard integer $n \in N$

$$\left| \| \sum r_j w_{n_j} \| - \| \sum r_j y_j \| \right| < 1/n .$$

(3) For every $j \in J_0$ we have $w_j = u_j$.

If such a family $\langle w_j | j \in J\rangle$ exists, then it is easy to show that the mapping T which takes each y_j to $\pi(w_j)$ $(j \in J)$ extends to the desired linear isometry on all of Y.

Now Theorem 7.1 asserts that for each finite system S of conditions from (1), (2), and (3), it is possible to choose w_j's satisfying S. Since the total number of conditions in (1), (2), and (3) is $< \kappa$, it follows that $\langle w_j | j \in J\rangle$ can be found satisfying all of them, because of the assumption that $^*M$ is κ-saturated.

Note that in view of our general assumption that $^*M$ is $\aleph_1$-saturated, the statement of Corollary 7.2 will always hold if X and Y are separable. In particular this means that the two spaces $\widehat{V'}$ and $(\hat{V})'$ have exactly the same separable subspaces. Recall that $\widehat{V'} = (\hat{V})'$ if and only if $\hat{V}$ is super reflexive. (Proposition 3.11.)

The special case of Theorem 7.1 where X and Y have dimension 1 was treated in [HMR 3]. This case follows from a simple application of the Hahn-Banach theorem inside $^*M$. The argument due to Stefan Heinrich reduces the general theorem to this special case, by regarding operators from Y into $\hat{V}'$ or into $\widehat{V'}$ as elements of appropriate Banach spaces.

First we give some elementary, standard facts about certain spaces of operators. For Banach spaces E and F, let $L(E,F)$ denote the space of all bounded linear operators from E into F, with the operator norm $\|\cdot\|$. If E is finite dimensional we also consider the space $N(E,F)$ of all bounded linear operators from E into F with a different norm $n(T)$ defined by

$$n(T) = \inf \text{ of sums } \sum_{j=1}^{k} \|\phi_j\| \cdot \|f_j\|$$
$$\text{where } \phi_1,\ldots,\phi_k \in E', \quad f_1,\ldots,f_k \in F \text{ and for all } x \in E \quad T(x) = \sum_{j=1}^{k} \phi_j(x) \cdot f_j .$$

It is not difficult to show that $n(T)$ is a norm. (Note that this definition of $n(T)$ makes sense whenever T has finite rank, even if its domain is infinite dimensional.) Next we show that $n(T)$ is equivalent to the operator norm:

Lemma 7.3. If E has dimension m and $T: E \to F$ is a bounded linear operator, then $\|T\| \le n(T) \le m\|T\|$.

Proof. The left inequality follows immediately from the triangle inequality. For the second, let $Y = T(E)$, so Y has dimension $k \le m$ and let $y_1,\ldots,y_k$ be a basis for Y with coordinate functionals $\psi_1,\ldots,\psi_k \in Y'$. We may assume that these were chosen so that $\|y_j\| = \|\psi_j\| = 1$ for all j (i.e., they form an Auerbach system for Y and Y'.) Use the Hahn-Banach theorem to extend each ψ_j to a norm 1 linear functional defined on all of F. Now define ϕ_j on E by $\phi_j(x) = \psi_j(T(x))$ for all $x \in E$. Each $\|\phi_j\|$ has norm $\le \|T\|$. Also

$$T(x) = \sum_{j=1}^{k} \psi_j(T(x))y_j$$

$$= \sum_{j=1}^{k} \phi_j(x)y_j \qquad (\text{all } x \in E)$$

showing that

$$n(T) \le \sum_{j=1}^{k} \|\phi_j\| \cdot \|y_j\| \le k\|T\| \le m\|T\|$$

as desired.

If E is any Banach space and $T: E \to E$ is any finite rank linear operator, recall that the trace of T is defined by

$$\operatorname{tr}(T) = \sum_{j=1}^{k} \phi_j(x_j)$$

where $\phi_1,\ldots,\phi_k \in E'$, $x_1,\ldots,x_k \in E$ and T is represented by

$$T(x) = \sum_{j=1}^{k} \phi_j(x) \cdot x_j$$

for all $x \in E$. (It is an elementary computation to show that this does not depend on the particular representation of T.)

It is convenient to introduce a standard "tensor product" notation for rank 1 linear operators from E to F: given $\phi \in E'$ and $f \in F$, $\phi \otimes f$ denotes the operator on E defined by

$$(\phi \otimes f)(x) = \phi(x) \cdot f$$

for all $x \in E$. An elementary calculation shows that

$$n(\phi \otimes f) = \|\phi \otimes f\| = \|\phi\| \cdot \|f\|$$

so that $n(\cdot)$ and the operator norm agree on rank 1 operators. In general however they do not agree.

Lemma 7.4. Let F be any Banach space and let E be finite dimensional. The dual space of $N(E,F)$ is $L(F,E)$, under the pairing

$$\langle T,S\rangle = \operatorname{tr}(ST)$$

for $T \in N(E,F)$ and $S \in L(F,E)$.

Proof. Fix $S \in L(F,E)$ and let ϕ_S be the linear functional on $N(E,F)$ defined by

$$\phi_S(T) = \operatorname{tr}(ST) .$$

We will show first that the norm of ϕ_S is exactly $\|S\|$.

Fix $\varepsilon > 0$ and $T \in N(E,F)$. Choose $\phi_1,\dots,\phi_k \in E'$, $f_1,\dots,f_k \in F$ so that $T(x) = \Sigma\, \phi_j(x) f_j$ for all $x \in E$ and so that $(1+\varepsilon) \cdot n(T) \geq \Sigma \|\phi_j\| \cdot \|f_j\|$. Then $\operatorname{tr}(ST) = \operatorname{tr}(TS) = \Sigma\, \phi_j(S(f_j))$. Hence

$$|\operatorname{tr}(ST)| \leq \Sigma \|\phi_j\| \cdot \|f_j\| \cdot \|S\|$$

$$\leq (1+\varepsilon) n(T) \|S\| .$$

Letting $\varepsilon \to 0$ we see that the linear functional ϕ_S on $N(E,F)$ has norm $\leq \|S\|$.

On the other hand, consider any rank 1 operator $\psi \otimes f$ in $N(E,F)$, so $\psi \in E'$ and $f \in F$. Recall that $n(\psi \otimes f) = \|\psi\| \cdot \|f\|$. Also for $T = \psi \otimes f$ we have

$$\langle T,S\rangle = \operatorname{tr}(ST) = \operatorname{tr}(TS) = \psi(S(f)) ,$$

and hence,

$$|\psi(S(t))| \leq \|\phi_S\| n(T) = \|\phi_S\| \cdot \|\psi\| \cdot \|f\| .$$

Taking the supremum over all ψ with $\|\psi\| \leq 1$ and then over all f with $\|f\| \leq 1$ gives

$$\|S\| \leq \|\phi_S\| .$$

Together these show $\|S\| = \|\phi_S\|$ for each $S \in L(F,E)$. Since the mapping $S \to \phi_S$ is evidently linear, we have shown that it gives an isometric embedding of $L(E,F)$ into the dual space of $N(E,F)$. It remains to show that this mapping is onto.

Let ϕ be any bounded linear functional on $N(E,F)$. For each $f \in F$ define a linear functional $S(f)$ on E' by setting $S(f)$ at $\psi = \phi(\psi \otimes f)$. Then S defined this way is a linear operator from F into $E'' = E$. (Recall E is finite dimensional.) Since $N(E,F)$ is the linear span of these rank 1 operators, it is easy to check from this definition that

$$\langle T,S\rangle = \phi(T)$$

holds for all $T \in N(E,F)$. This completes the proof.

Along with these standard elementary facts about operators, we need two easy observations about nonstandard hulls of operator spaces. If V,W are internal Banach spaces, we may consider the internal spaces $L(V,W)$ and $N(V,W)$. We will denote their nonstandard hulls by $\hat{L}(V,W)$ and $\hat{N}(V,W)$ respectively.

<u>Lemma 7.5.</u> Let E be a standard, finite dimensional Banach space and let W be any internal Banach space in $^*M$. Then

1) $\hat{L}(^*E,W) = L(E,\hat{W})$
2) $\hat{N}(^*E,W) = N(E,\hat{W})$.

<u>Proof</u>: In each case, the "equality" means that there is a natural identification of the nonstandard hull on the left with the standard space on the right (with norms included.)

(1) Given any internal linear operator T from $^*E$ into W, if the norm of T is finite then we may use T to obtain an operator $\hat{T}: E \to \hat{W}$. (Namely, $\hat{T}(\pi(x))$ is defined to be the element of W determined by $T(x)$.) It is immediate that $\|\hat{T}\| = st\|T\|$. Moreover, since E is finite dimensional, every linear operator from E into $\hat{W}$ arises in this way.

(2) As in (1), we identify $\hat{N}(^*E,W)$ and $N(E,\hat{W})$ as sets via the mapping $T \to \hat{T}$ from finite elements of $N(^*E,W)$ to elements of $N(E,\hat{W})$. Lemma 7.3 shows that for any internal linear map $T: {}^*E \to W$, $\|T\|$ is finite iff $n(T)$ is finite and $\|T\|$ is infinitesimal iff $n(T)$ is infinitesimal. This means that we may regard $\hat{L}(^*E,W)$ and $\hat{N}(^*E,W)$ as being equal as sets. It remains to show that if $T: {}^*E \to W$ is an internal linear map and $n(T)$ is finite, then $n(\hat{T}) =$

st(n(T)).

It is easy to see that $\mathrm{st}(n(T)) \leq n(\hat{T})$, since any representation of $\hat{T}$ gives rise to one for T: If $\phi_1,\ldots,\phi_k \in E'$ and $f_1,\ldots,f_k \in \hat{W}$ satisfy

$$\hat{T}(x) = \Sigma\, \phi_j(x) \cdot f_j$$

then we choose $w_1,\ldots,w_k \in W$ with $\pi(w_j) = f_j$ for $j = 1,\ldots,k$ and define S on *E by

$$S(x) = \Sigma\, {}^*\phi_j(x)\, w_j \, .$$

Then for all standard $x \in E$ we easily see that $S(x)$ and $T(x)$ are infinitely close. Hence $\|S - T\|$ and $n(S - T)$ are both infinitesimal, since E is a standard finite dimensional space. Also we see

$$\begin{aligned} \mathrm{st}(n(S)) &\leq \mathrm{st}(\textstyle\sum \|\phi_j\| \cdot \|w_j\|) \\ &\leq \textstyle\sum \|\phi_j\| \cdot \|f_j\| \, . \end{aligned}$$

Now taking the inf over all such representations of $\hat{T}$ yields

$$\mathrm{st}(n(T)) = \mathrm{st}(n(S)) \leq n(\hat{T}).$$

Now we prove the opposite inequality. Let $\phi_1,\ldots,\phi_\omega$ and $w_1, \ldots,w_\omega$ be internal sequences, in *E' and W respectively, so that for all $x \in {}^*E$

$$T(x) = \sum_{j=1}^{\omega} \phi_j(x)\, w_j \, .$$

(Here $\omega \in {}^*\mathbb{N}$ and ω may be infinite.) We may choose this representation of T so that

$$n(T) \approx \sum_{j=1}^{\omega} \|\phi_j\| \cdot \|w_j\| \, .$$

for convenience we may assume $\|\phi_j\| = 1$ for all j, $1 \leq j \leq \omega$.

Fix $\varepsilon > 0$ (ε standard). Since E is a standard, finite dimensional space, there is a list of linear functionals $\psi_1,\psi_2,\ldots,\psi_k$ in E', all of norm 1, such that every norm 1 linear functional is within ε (in norm) of some ψ_i. (Here k is a standard integer.) This also holds in *E' and so we can select an internal sequence $n_1,n_2,\ldots,n_\omega$ of integers (each n_j satisfying $1 \leq n_j \leq k$) so that $1 \leq j \leq \omega$ implies $\|\phi_j - {}^*\psi_{n_j}\| \leq \varepsilon$. Define $S\colon {}^*E \to W$ by

$$S(x) = \sum_{j=1}^{\omega} {}^*\psi_{n_j}(x) \cdot w_j .$$

We see that $n(T - S)$ is bounded above by

$$\sum_{j=1}^{\omega} \| \phi_j - {}^*\psi_{n_j}(x) \| \cdot \| w_j \| \le \varepsilon \sum \| w_j \| \le 2 \in n(T).$$

Hence

$$n(\hat{T} - \hat{S}) \le \dim(E) \cdot \| \hat{T} - \hat{S} \|$$

$$\approx \dim(E) \| T - S \| \le \dim(E) n(T - S)$$

$$\le 2\varepsilon \cdot \dim(E) \cdot n(T).$$

Moreover, by setting $v_i = \sum_{n_j = i} w_j \quad (i = 1, 2, \ldots, k)$ we can represen S as a finite sum

$$S(x) = \sum_{i=1}^{k} {}^*\psi_i(x) \cdot V_i .$$

Therefore

$$\hat{S}(x) = \sum_{i=1}^{k} \psi_i(x) \cdot z_i$$

for all $x \in E$, where $z_i = \pi(v_i) \in \hat{W}$ for $1 \le i \le k$. Then we see

$$n(\hat{S}) \le \sum_{i=1}^{k} \| \psi_i \| \| z_i \| = \sum_{i=1}^{k} \| z_i \|$$

$$\le \sum_{i=1}^{k} \| v_i \| + \varepsilon$$

$$\le \sum_{j=1}^{\omega} \| w_j \| + \varepsilon \le n(T) + 2\varepsilon .$$

Hence

$$n(\hat{T}) \le n(\hat{T} - \hat{S}) + n(\hat{S})$$

$$\le 2\varepsilon \cdot \dim(E) \cdot n(T) + n(T) + 2\varepsilon.$$

Letting $\varepsilon \to 0$ (ε standard) yields

$$n(\hat{T}) \le st(n(T)),$$

and completes the proof.

<u>Proof of Theorem 7.1.</u> Let X, Y, V, V' be as in the statement of

Theorem 7.1. We consider Y as a standard finite dimensional space and let $J: Y \to (\hat{V})'$ be the inclusion mapping, so $\|J\| = 1$. Then J induces a linear functional ϕ on $N(Y',\hat{V})$ by the definition

$$\phi(T) = tr(T' \cdot J).$$

(Here $T': \hat{V}' \to Y$ is the adjoint mapping of $T: Y' \to \hat{V}$.) An elementary calculation shows that for any T

$$|\phi(T)| = |tr(T' \cdot J)| \leq n(T)\|J\| = n(T) .$$

Indeed, if T is represented in the form

$$T(\psi) = \sum_{j=1}^{n} \psi(y_j) \cdot f_j$$

for $y_1,\ldots,y_n \in Y$ and $f_1,\ldots,f_j \in \hat{V}$, then $T' \cdot J$ is given on $(\hat{V})'$ by

$$T'(\psi) = \sum_{j=1}^{n} \psi(f_j) \cdot y_j$$

and hence

$$tr(T' \cdot J) = \Sigma \langle J(y_j),f_j\rangle .$$

Therefore $|tr(T' \cdot J)| \leq \sum_{j=1}^{n} \|y_j\| \cdot \|f_j\| \cdot \|J\|$. Taking the inf over all representations of T leads to the desired inequality

$$|tr(T' \cdot J)| \leq n(T)\|J\| = n(T).$$

This means that ϕ has norm ≤ 1 as a linear functional on $N(Y',\hat{V})$.

Now let Z be the internal Banach space $N(^*Y',V)$. By Lemma 7.5 $\hat{Z}$ is $N(Y',\hat{V})$ so that by Lemma 7.4, $(\hat{Z})'$ is $L(\hat{V},Y')$. On the other hand, if we apply 7.4 internally, we see that the internal dual space Z' is $L(V,{}^*Y')$.

Next we observe that $L(\hat{V},Y')$ is linearly isometric to $L(Y,(\hat{V})')$, via the adjoint operation. The same idea applied internally shows that $L(V,{}^*Y')$ is linearly isometric (by the <u>internal</u> adjoint operation) to $L({}^*Y,V')$. (Recall that V' is the <u>internal</u> dual space of V.) Therefore by Lemma 7.4 we see that $\widehat{Z'}$ may be regarded as being $\hat{L}(Y,\widehat{V'})$.

Thus we have the following diagram

$$\begin{array}{ccl} & & \phi \in (\hat{Z})' = L(Y,(\hat{V})') \\ N(Y',\hat{V}) = \hat{Z} & \begin{array}{c} \nearrow \\ \searrow \end{array} & \quad\;\; \cup\!| \\ & & \widehat{Z'} = L(Y,\widehat{V'}). \end{array}$$

It is easy to check that the pairing between $\hat{Z}$ and $(\hat{Z})'$ becomes, under this identification, $\langle T,S\rangle = \mathrm{tr}(T'S)$, where $T \in N(Y',\hat{V})$ and $S \in L(Y,(\hat{V})')$.

Now let $Y_1 = Y \cap \widehat{V'}$ and write $Y = Y_1 \oplus Y_2$ as a direct sum, for some Y_2. Let W denote the internal linear subspace of $N(*Y',V)$ which consists of all $T\colon *Y' \to V$ which can be represented as a $*$-finite sum

$$T(\psi) = \sum_{j=1}^{\omega} \psi(y_j) \cdot w_j$$

for some internal sequences $y_1,\ldots,y_\omega$ from $*Y_1$ and $w_1,\ldots,w_\omega$ from V. (Note that $*Y'' = *Y = *Y_1 \oplus *Y_2$, so our restriction on T is simply that the linear functionals which go into its representation all come from $*Y_1$.)

Since $Y_1 \subseteq \widehat{V'}$ and Y_1 has standardly finite dimension, we can obtain an internal linear functional χ on W such that $\hat{\chi}$ agrees with ϕ on $\hat{W} \subseteq \hat{N}(*Y',V) = N(Y',\hat{V})$. (If $J_1\colon Y \to \widehat{V'} \subseteq \hat{V}'$ is defined to be 0 on Y_2 and to be the inclusion mapping on Y_1, then $\hat{\chi}$ corresponds to J_1 in the same way that ϕ corresponds to J.) Since $N(Y',X) \subseteq N(Y',\hat{V})$ and $N(Y',X)$ is separable, we can combine an $\aleph_1$-saturation argument together with an internal application of the Hahn-Banach theorem to obtain an internal linear functional η defined on all of $N(*Y',V)$ such that $\hat{\eta}$ agrees with ϕ on $\hat{W}$ and on $N(Y',X)$ and such that $\hat{\eta}$ still has norm ≤ 1. But then η is in the internal dual of $N(*Y',V)$, which is $L(V,*Y')$, here being identified with $L(*Y,V')$ by the adjoint operation. Therefore $\hat{\eta}$ corresponds to an element of $\hat{L}(*Y,V') = L(Y,\widehat{V'})$, that is, to an operator $\hat{S}\colon Y \to \widehat{V'}$. (Here $S\colon *Y \to V'$ corresponds to η.) It is now possible to show that $\hat{S}$ is the desired operator embedding Y into $\widehat{V'}$. The fact that $\|\hat{S}\| \leq 1$ is a consequence of $\|\eta\| \leq 1$. Clause (a) of Theorem 7.1 follows by considering the action of ϕ (and hence also of $\hat{\eta}$) on $N(Y',X)$. Clause (b) follows by using the action of ϕ (and hence also of $\hat{\eta}$) on $\hat{W}$. Finally we note that $\|\hat{S}\| \geq 1$ follows from Clause (a), as long as X is taken large enough to norm all the linear functionals in Y. (This may require expanding X to an infinite dimensional but separable subspace of $\hat{V}$. The only fact about X used above was that $N(Y',X)$ is separable, and this is still true if X is separable, since Y' is finite dimensional.) This completes the proof of Theorem 7.1.

Next we consider a useful technical tool for treating finite rank operators on nonstandard hulls. It is an immediate consequence of

Theorem 7.1 and Corollary 7.2 and leads to the most important applications of the Kürsten-Stern local duality result.

Theorem 7.6. Assume the nonstandard superstructure $*M$ is κ-saturated ($\kappa \geq \aleph_1$.) Let V be an internal Banach space. Suppose $X \subseteq \hat{V}$ is a subspace of topological density character $< \kappa$ and suppose $T: \hat{V} \to \hat{V}$ is a finite rank bounded linear operator.

Then there exists an internal linear operator $S: V \to V$ such that the operator $\hat{S}: \hat{V} \to \hat{V}$ has the same range as T, has the same norm as T and agrees with T on X. If T is a projection then we may take $\hat{S}$ to be a projection too.

Proof. Since T has finite rank, there exist $x_1,\ldots,x_m \in \hat{V}$ and $\phi_1,\ldots,\phi_m \in (\hat{V})'$ so that

$$T(x) = \sum_{j=1}^{m} \phi_j(x) \cdot x_j$$

for all $x \in \hat{V}$. (Note that m is a standard integer, of course.) Choose $v_1,\ldots,v_m \in V$ so that $x_j = \pi(v_j)$ for each $j = 1,\ldots,m$. Now we use Corollary 7.2 to obtain linear functionals $\psi_1,\ldots,\psi_m \in \widehat{V'}$ so that (a) the linear mapping from $\operatorname{span}(\phi_1,\ldots,\phi_m)$ to $\operatorname{span}(\psi_1,\ldots,\psi_m)$, taking ϕ_j to ψ_j for each j, is an isometry; (b) ψ_j and ϕ_j agree on X for each j. Choose internal linear functionals $\chi_1,\ldots,\chi_m$ in V' so that $\pi(\chi_j) = \psi_j$ for each j. Define $S: V \to V$ by

$$S(v) = \sum_{j=1}^{m} \chi_j(v) \cdot v_j$$

so that S is an internal linear operator. It is easy to see that $\hat{S}: \hat{V} \to \hat{V}$ is in fact given by

$$\hat{S}(x) = \sum_{j=1}^{m} \psi_j(x) \cdot x_j$$

for all $x \in \hat{V}$.

It is immediate from this that $\hat{S}$ agrees with T on X. If X is enlarged to include the range of T and to norm the range of T, then this implies $\|T\| \leq \|\hat{S}\|$ and that $\hat{S}$ is a projection if T is one. The fact that $\|\hat{S}\| \leq \|T\|$ comes from an easy calculation, using the isometric equivalence between the sequences of linear functionals $(\phi_1,\ldots,\phi_m)$ and $(\psi_1,\ldots,\psi_m)$ used in defining T and $\hat{S}$.

The typical way in which Theorem 7.6 is applied to a standard Banach space E goes as follows: one has a finite rank operator T

on the nonstandard hull $\hat{E}$ and T is known to have some nice properties. For example it may be a projection, its range may be λ-isomorphic to some specific space such as $\ell_p(m)$, it may be close to the identity on a finite dimensional subspace of E, etc. Using Theorem 7.6 we may replace T by $\hat{S}$, where S is an internal linear operator on $^*E$, without disturbing any of these nice properties. Then a straightforward transfer argument leads to the existence of a finite rank operator U on E itself such that U also shares the same nice properties possessed by the original operator T.

Stefan Heinrich [HEI 1] has used these methods to prove a number of interesting results about the Uniform Approximation Property, answering questions posed by Lindenstrauss and Tzafriri [LT 1]. See [HEI 4, Section 9] for a good exposition of Heinrich's proofs. He shows:

Theorem 7.7. (S. Heinrich) Let E be a standard Banach space.

(i) E has the uniform approximation property if and only if the dual space E' has the uniform approximation property.

(ii) If E has the uniform approximation property then the same is true of $L_p(E)$ for $1 \leq p \leq \infty$.

(iii) If E is super-reflexive and has the uniform approximation property, then E has the $(1+\varepsilon)$-uniform approximation property for every $\varepsilon > 0$.

The results (i) and (ii) were first proved by S. Heinrich using these methods. Part (iii) had been proved earlier by Lindenstrauss and Tzafriri using more direct methods. Heinrich's proof of (iii) is much easier and gives a good example of how the methods discussed above based on Theorem 7.6, are used:

Suppose E satisfies the hypotheses in Theorem 7.7(iii). Then $\hat{E}$ is reflexive and it is easy to show that $\hat{E}$ also has the uniform approximation property, by a transfer argument. A well-known theorem due to Grothendieck implies that $\hat{E}$ has the metric approximation property. Now the machinery based on Theorem 7.6 comes into play: one uses it to show that because $\hat{E}$ has the 1-bounded approximation property, then E must have the $(1+\varepsilon)$-uniform approximation property for each $\varepsilon > 0$. Given $x_1,\ldots,x_n \in E \subseteq \hat{E}$, we know that there is a finite rank operator T on $\hat{E}$ such that $\|T(x_j) - x_j\| < \varepsilon$ for $j = 1,2,\ldots,n$. Moreover, because this corresponds essentially to an internal condition, the rank of T can be uniformly bounded as a function of n,ε and the norm of T. The approach based on Theorem 7.6 then allows one to deduce the existence of such an operator $T: E \to E$, as desired.

Another standard result proved first using these methods by Henson is the following. It settles positively a conjecture made by Enflo and Rosenthal [ER].

Theorem 7.8. (Henson) Let $1 \le p < 2$ and μ any measure. For each $\lambda \ge 1$, $\varepsilon > 0$ and $n \in \mathbb{N}$ there exists $m = m(\lambda, \varepsilon, n) \in \mathbb{N}$ with the following property: if $x_1, \ldots, x_m$ have norm 1 and are in $L_p(\mu)$ and if $(x_1, \ldots, x_m)$ is λ-equivalent to the usual basis of $\ell_p(m)$, then there exist $i_1, \ldots, i_n$ such that the span of $(x_{i_1}, \ldots, x_{i_n})$ is the range of a projection P on $L_p(\mu)$ whose norm is $\le \lambda + \varepsilon$.

For a proof of this result see [HEN 4] or [HEI 4, Section 10].

Other uses of Theorem 7.6 give information about the occurrence of uniformly complemented families of finite dimensional subspaces of E. We give a few examples:

Corollary 7.9. Let V be an internal Banach space and let $1 \le p < \infty$ $\lambda \ge 1$, and $\eta \ge 1$ be standard numbers. Then the following are equivalent:

1) $\hat{V}$ has a subspace X which is λ-isomorphic to ℓ_p and is the range of a norm $\le \eta$ projection on $\hat{V}$.
2) For each $n \in \mathbb{N}$, $\hat{V}$ has a subspace which is λ-isomorphic to $\ell_p(n)$ and is the range of a norm $\le \eta$ projection on $\hat{V}$.

Proof. (1) => (2) is trivial, since there are norm 1 projections from ℓ_p onto $\ell_p(n)$.

(2) => (1): Using Theorem 7.6 we may assume the projection of $\hat{V}$ onto the $\ell_p(n)$ subspace is $\hat{S}_n$ where S_n is internal. For a properly chosen infinite $\omega \in {}^*\mathbb{N}$, $\hat{S}_\omega$ is a projection of $\hat{V}$ onto a subspace of $\hat{V}$ which is λ-isomorphic to $\hat{\ell}_p(\omega)$ and which has norm $\le \eta$. The proof is completed by using the fact that $\hat{\ell}_p(\omega)$ is linearly isometric to $L_p(\mu)$ for a measure μ and has a norm 1 complemented subspace which is linearly isometric to ℓ_p.

When V is ${}^*E$ for a standard Banach space E, we get the following interesting consequence.

Corollary 7.10. Fix $1 \le p < \infty$. Then E has uniformly complemented subspaces uniformly isomorphic to $\{\ell_p(n)\}$ if and only if $\hat{E}$ has a complemented subspace which is isomorphic to ℓ_p.

Proof. The direction from E to $\hat{E}$ is trivial, using the fact that

ℓ_p is complemented in each $\hat{\ell}_p(\omega)$. For the converse, use Theorem 7.6 to obtain a family of internal S_n, as in the proof of Corollary 7.9. A perturbation argument permits one to assume that the $\{S_n\}$ are actually projections on $^*E$ whose ranges are uniformly isomorphic to $\{\ell_p(n)\}$ and whose norms are uniformly bounded (by a standard bound). A straightforward transfer argument pulls the existence of such S_n from $^*E$ back to E.

Many other applications of the Kürsten-Stern local duality result, to operator ideals, to finitely representing operators, etc., have been found by Stefan Heinrich. See [HEI 1], [HEI 3] especially.

We end this Section with a useful application of the Kürsten-Stern theorem to the existence of nice finitely represented spaces.

Lemma 7.11. Suppose E, F are standard Banach spaces and E is finitely represented in F. There exists a Banach space X such that $E \subseteq X$, X is finitely represented in F and X' is finitely represented in F'. The space X may be taken to have the same density character as E.

Proof. We may regard E as a subspace of the nonstandard hull $\hat{F}$. (See Theorem 3.2 and the remarks following it.) Obtain $X \subseteq \hat{F}$ with the same density character as E and satisfying $E \subseteq X \prec_A \hat{F}$, using the Downward Löwenheim-Skolem Theorem (Theorem 5.6). In particular, X is a reflecting subspace of $\hat{F}$ and so X' embeds isometrically in $(\hat{F})'$ by Corollary 3.15. It follows from the Kursten-Sterm local duality theorem that $(\hat{F})'$ is finitely represented in $\widehat{F'}$ and therefore X' is finitely represented in F', as desired.

A similar argument has been used by Heinrich and Mankiewicz [HMK] to prove that every dual space Y' has a complemented subspace of cardinality at most the continuum. To do this use the Downward Löwenheim-Skolem Theorem (Theorem 5.6) to get a separable X such that $X \prec_A Y$. Then Corollary 3.15 yields that X', whose cardinality is at most the continuum, embeds as a 1-complemented subspace of Y'.

Section 8. Back and Forth.

In this Section we discuss some tools for understanding and dealing with the equivalence relation $\equiv_A$. These tools all involve families of linear isomorphisms between subspaces of Banach spaces and they derive from the so-called back-and-forth arguments in model theory. (See the expository article by J. Barwise [BA] for example.)

First we consider a new game-theoretic concept of equivalence between Banach spaces. These games are motivated by the Ehrenfeucht-Fraissé games which arise in logic. [EH] [FR].

Fix an integer k and a number $\lambda \geq 1$. Let X and Y be Banach spaces. The <u>(k,λ)-game between X and Y</u> is a k-step game for two players I and II, who alternate their moves. The steps of the game generate a sequence $T_0, T_1, \ldots, T_k$, where each T_j is a non-singular linear mapping from a $\leq$ j-dimensional subspace of X into Y. The initial map T_0 is trivial, defined only on $\{0\}$. At the (j+1)st play, I chooses an arbitrary element of X or of Y. Then player II chooses the map T_{j+1} so that it extends T_j and so that its domain or range (as appropriate) contains the element just chosen by I.

The game is won by II if and only if the mapping T_k satisfies $\|T_k\| \leq \lambda$ and $\|T_k^{-1}\| \leq \lambda$. Otherwise I wins.

Since this game has a fixed finite number of steps, it is determined: either player I or player II has a winning strategy. The assertion that player II has a winning strategy for the (k,λ)-game between X and Y expresses a certain local geometric equivalence between X and Y for subspaces of dimension $\leq k$.

The following characterization of $\equiv_A$ in terms of these games will be proved in [HH]. (Compare to Proposition 5.1.)

<u>Theorem 8.1.</u> For any Banach spaces X and Y, the following are equivalent

(i) $X \equiv_A Y$

(ii) for every integer k and every $\lambda > 1$, player II has the winning strategy in the (k,λ)-game.

For fixed k and λ, to say that player II has a winning strategy in the (k,λ)-game between X and Y means intuitively that X and Y are locally λ-isomorphic, in respect to the type and relative placement of their $\leq k$ dimensional subspaces.

One ingredient in the proof of Theorem 8.1 is the following, which gives another relation between an internal Banach space V and its nonstandard hull $\hat{V}$.

<u>Lemma 8.2.</u> Let V, W be internal Banach spaces. Fix a standard integer k and standard real numbers $\lambda > \eta \geq 1$.

(a) If player II has a winning strategy in the (k,η)-game between $\hat{V}$ and $\hat{W}$, then player II has an internal winning strategy in the internal (k,λ) game between V and W.

(b) If player II has an internal winning strategy in the internal (k,η)-game between V and W, then player II has a winning strategy in the (k,η)-game between $\hat{V}$ and $\hat{W}$.

Roughly speaking, these results are proved by playing the (k,η)-game in V, W at the same time as in $\hat{V}, \hat{W}$, and arranging that if $S_0,\ldots,S_k$ is the play in V,W and $T_0,\ldots,T_k$ is the play in $\hat{V}, \hat{W}$, then $T_j = \hat{S}_j$ for $j = 0,1,\ldots,k$.

These results provide a tool for obtaining standard results about Banach spaces, phrased in this game-theoretic terminology, to be obtained directly from isometric classification of nonstandard hulls. As an example, we give an application of results from Section 9 (and [HMR 4]) to the study of the asymptotic local structure of the finite dimensional sequence spaces $\ell_p(n)$, as p and n tend to ∞. This result shows that this asymptotic local structure depends only on the limit of $n^{1/p}$.

<u>Theorem 8.3.</u> (i) Fix an integer k and a real number $\lambda > 1$. There exists an integer $M = M(k,\lambda)$ such that if $m,n,p,q \geq M$ and $|m^{-1/p} - n^{-1/q}| \leq 1/M$ then player II has a winning strategy in the (k,λ)-game between $\ell_p(m)$ and $\ell_q(n)$.

(ii) On the other hand, given $\varepsilon > 0$ there are integers M,k and a number $\lambda > 1$ (all depending only on ε) such that if $m,n,p,q \geq M$ and $|m^{-1/p} - n^{-1/q}| \geq \varepsilon$, then player I has a winning strategy in the (k,λ)-game between $\ell_p(m)$ and $\ell_q(n)$.

<u>Sketch of Proof of (i)</u>: Fix k and λ. Passing to a nonstandard superstructure $^*M$, it suffices to show that the $*$-transform of the assertion in (i) holds for every infinite integer M in $^*\mathbb{N}$. So let M be infinite and take $m,n,p,q \geq M$ (with $m,n \in {}^*\mathbb{N}$ and $p,q \in {}^*\mathbb{R}$.) If $|m^{-1/p} - n^{-1/q}| \leq 1/M$ then it follows that either $m^{1/p}$ and $n^{1/q}$ are both infinite or they are both finite and have the same standard part. Using Theorem 9.5 we see that the nonstandard hulls $\hat{\ell}_p(m)$ and $\hat{\ell}_q(n)$ are linearly isometric. It is a trivial consequence of this that player II has a winning strategy in the $(k,1)$-game between $\hat{\ell}_p(m)$ and $\hat{\ell}_q(n)$. By Lemma 8.2 we get that player II has an internal winning strategy in the internal (k,λ)-game between $\ell_p(m)$ and $\ell_q(n)$. This is exactly what was supposed to be proved.

In great generality, classification results (under isometry or under isomorphism) for classes of nonstandard hulls can be used to obtain standard theorems similar to Theorem 8.3. For example, Theorem 6.7 yields the following result which is an isomorphic version of Theorem 8.3(i).

<u>Theorem 8.4.</u> There exists a real number $\lambda_o > 1$ with the following property: for each integer k there exists an integer $M = M(k)$ such

that if $m,n,p,q \geq M$, then player II has a winning strategy in the (k,λ_0)-game between $\ell_p(m)$ and $\ell_q(n)$.

Note that Theorem 8.3(ii) shows that λ_o is bounded away from 1.

In the rest of this Section we discuss a useful tool for proving that the nonstandard hulls $\hat{V}$, $\hat{W}$ (of two internal Banach spaces V, W) are linearly isometric (Theorem 8.7). The use of this result will be illustrated in Section 9.

Let E and F be Banach spaces and let $\mathcal{F}$ be a nonempty family of linear isometries T from a subspace $\mathrm{dom}(T)$ of E into a subspace $\mathrm{ran}(T)$ of F.

Definition 8.5. The family $\mathcal{F}$ is said to be an extendable family of isometries if for each T in $\mathcal{F}$, each $\varepsilon > 0$, each x in E and each y in F there exists S in $\mathcal{F}$ so that:

1) $\mathrm{dist}(x,\mathrm{dom}(S)) \leq \varepsilon\|x\|$,
2) $\mathrm{dist}(y,\mathrm{ran}(S)) \leq \varepsilon\|y\|$, and
3) for every z in $\mathrm{dom}(T)$ there exists z' in $\mathrm{dom}(S)$ with $\|z - z'\| \leq \varepsilon\|z\|$ and $\|Tz - Sz'\| \leq \varepsilon\|z\|$.

Lemma 8.6. Let $\mathcal{F}$ be an extendable family of isometries of subspaces of E into F. Then given T in $\mathcal{F}$, $\varepsilon > 0$, x in E with $\|x\| \leq 1$ and y in F with $\|y\| \leq 1$ there exists S in $\mathcal{F}$, x'' in $\mathrm{dom}(S)$ and y'' in $\mathrm{ran}(S)$ such that:

1') $\|x''\| \leq 1$ and $\|x - x''\| \leq \varepsilon$,
2') $\|y''\| \leq 1$ and $\|y - y''\| \leq \varepsilon$,
3') for every z in $\mathrm{dom}(T)$ with $\|z\| \leq 1$ there exists z'' in $\mathrm{dom}(S)$ with $\|z''\| \leq 1$, $\|z - z''\| \leq \varepsilon$ and $\|Tz - Sz''\| \leq \varepsilon$.

Proof. Applying the definition of an extendable family with ε replaced by $\varepsilon/2$, we obtain x' in $\mathrm{dom}(S)$ with $\|x - x'\| \leq \varepsilon/2$. It follows that $\|x'\| \leq 1 + \varepsilon/2$, so some scalar multiple x'' of x' has norm one and satisfies $\|x'' - x'\| \leq \varepsilon/2$. Thus (1') holds and (2') and (3') are established similarly.

The following perturbation result for positive bounded formulas is a basic element of the model theory of Banach spaces. It will be proved in [HMR 4]. (See also [HEN 5].)

Proposition 8.7. Let $\sigma(x_1,x_2,\ldots,x_n)$ be a positive bounded formula and let k be a positive integer. There exists $\varepsilon = \varepsilon(\sigma,k) > 0$ such that:

for any Banach space X, any n elements $x_1, x_2, \ldots, x_n$ of the closed unit ball of X and any $y_1, y_2, \ldots, y_n$ in X satisfying $\|x_i - y_i\| \leq \varepsilon$ for $1 \leq i \leq n$, we have

$$X \models \sigma_{k+1}(x_1, x_2, \ldots, x_n) \text{ implies } X \models \sigma_k(y_1, y_2, \ldots, y_n).$$

<u>Theorem 8.8.</u> Let E and F be Banach spaces. If there exists an extendable family F of isometries of subspaces of E into F, then $E \equiv_A F$.

<u>Proof.</u> We prove by induction on the complexity of positive bounded formulas that:

(*) If T belongs to F and $x_1, x_2, \ldots, x_n$ are in the closed unit ball of $\mathrm{dom}(T)$ then $E \models_A \sigma(\underline{x})$ if and only if $F \models_A \sigma(T\underline{x})$.

Then if σ is a sentence we have $E \models_A \sigma$ if and only if $F \models_A \sigma$, i.e., $E \equiv_A F$.

If σ is quantifier free the result follows since T is an isometry. Moreover the induction step is trivial when σ is $\tau \vee \gamma$ or $\tau \wedge \gamma$. Now consider the case when σ is $\exists_B z\, \tau(z, \underline{x})$ and we have $E \models_A \sigma(\underline{x})$. (By the symmetry of the conditions on F, it is enough here, and in the universal quantifier case, to show the implication from E to F.) Fix a positive integer k and select $\varepsilon > 0$ so that ε is less than both $\varepsilon(\tau, k+2)$ and $\varepsilon(\tau, k)$. (See Proposition 8.7.)

Now we are assuming we have z in E with $\|z\| \leq 1$ and $E \models \tau_{k+3}(z, \underline{x})$. By Lemma 8.6 we can find S in F and $w, y_1, y_2, \ldots, y_n$ in $\mathrm{dom}(S)$ so that $\|w\| \leq 1$ and $\|z - w\| \leq \varepsilon$ and, for each i, $\|x_i - y_i\| \leq \varepsilon$ and $\|Tx_i - Sy_i\| \leq \varepsilon$. Since $\varepsilon < \varepsilon(\tau, k+2)$ we have that $E \models \tau_{k+2}(w, \underline{y})$. Then by the induction hypothesis we have $F \models_A \tau_{k+2}(Sw, S\underline{y})$ and so $F \models \tau_{k+1}(Sw, S\underline{y})$. Now since $\varepsilon < \varepsilon(\tau, k)$ and $\|Sw\| \leq 1$, $\|Sy_i\| \leq 1$ for $1 \leq i \leq n$, by Proposition 8.7 $F \models \tau_k(Sw, T\underline{x})$ and hence $F \models (\exists_B z\, \tau(z, T\underline{x}))_k$. Since k was arbitrary, we have $F \models_A \exists_B z\, \tau(z, T\underline{x})$. The induction step for universal quantifiers can be handled similarly and we omit the details.

We note that if E and F are nonstandard hulls $\hat{V}$, $\hat{W}$, then the converse of Theorem 8.8 is true: if $\hat{V} \equiv_A \hat{W}$ then there is an extendable family F of isometries from subspaces of $\hat{V}$ into $\hat{W}$. (Note that when $*M$ has the $\aleph_0$-isomorphism property this fact is trivial, since then $\hat{V} \equiv_A \hat{W}$ implies $\hat{V}$ and $\hat{W}$ are isometric.)

Section 9. Hyperfinite dimensional Banach spaces.

Recall that if W is a *-finite dimensional space, then $\hat{W}$ is called a hyperfinite dimensional Banach space. For example we have the spaces $\hat{\ell}_p(n)$ where $n \in {}^*N$ and either $p \in {}^*\mathbb{R}$ with $p \geq 1$ or $p = \infty$. For each such hyperfinite dimensional space $\hat{W}$, we know by Corollary 5.7 that there exists a separable Banach space E such that $\hat{E}$ is linearly isometric to $\hat{W}$. We consider now what is known in the opposite direction.

Definition 9.1. A Banach space E is said to be approximately finite dimensional if a nonstandard hull $\hat{E}$ of E, formed with respect to an $\aleph_1$-saturated extension is linearly isometric to a hyperfinite dimensional Banach space.

The definition of approximately finite dimensional Banach space may be recast in term of approximations to positive bounded sentences [HEN 3].

Proposition 9.2. A Banach space E is approximately finite dimensional if and only if for every positive bounded sentence σ which is approximately true in E and for every k in N there is a finite-dimensional Banach space in which the approximation σ_k of σ holds.

Before we begin our discussion of which infinite dimensional spaces are approximately finite dimensional, we recall several consequences of the $\aleph_0$-isomorphism property [HEN 1].

Proposition 9.3. Assume ${}^*M$ possesses the $\aleph_0$-isomorphism property.

(i) Every two (externally) infinite internal subsets of ${}^*M$ have the same (external) cardinality.

(ii) Let A and B be two infinite internal sets in ${}^*M$ then there are bijections f and g of A onto B such that for a subset S of A:

S is internal if and only if $f(S)$ is internal if and only if either $g(S)$ or its complement in B is *-finite.

Of course in general the bijections guaranteed by Proposition 9.3 are not internal.

Using 9.3(i) it is easy to see that every Hilbert space H is approximately finite dimensional. Indeed if ${}^*M$ has the $\aleph_0$-isomorphism property, then $\hat{H}$ is isometric to $\hat{\ell}_2(n)$ for every infinite positive integer n. It is also easy to see that $\hat{\ell}_\infty$ is isometric to $\hat{\ell}_\infty(n)$ for any infinite n. Let f be a bijection of ${}^*N$ onto

$\{1,2,\ldots,n\}$ such that S is internal if and only if $f(S)$ is internal. Then one can define the isometry Φ of $\hat{\ell}_\infty$ onto $\hat{\ell}_\infty(n)$ by setting $\Phi(\chi_S) = \chi_{f(S)}$ for all internal subsets of $*N$, extending Φ by linearity to the dense subspace spanned by these characteristic functions and taking the unique extension to all of $\hat{\ell}_\infty$. If $1 \leq p < \infty$ then the general discussion of nonstandard hulls which are L_p-spaces [HEN 3], shows that $\hat{\ell}_p$ is isometric to $\hat{\ell}_p(n)$ for all infinite integers n.

Somewhat surprisingly, the sequence space c_0 is not approximately finite dimensional. It is easy to see that c_0 is not isometric to $\hat{\ell}_\infty$ (nothing can map onto χ_{*N}) or $\hat{\ell}_\infty(n)$ for any n. Thus there are no "natural" hyperfinite dimensional spaces which are candidates to be isometric to $\hat{c}_0$. However the argument that no such hyperfinite dimensional space can be found is rather intricate and involves the description of the Banach lattice structure of $\hat{c}_0$ in terms of the norm and the fact that with respect to this order structure every positive element of $\hat{c}_0$ is the supremum of the atoms it dominates [MO 2].

Aside from the spaces ℓ_p, $1 \leq p < \infty$, the only other large class of Banach spaces where it is known which ones are approximately finite dimensional is the class of spaces $C(K)$ with K compact. The following result is proved in [MO 3].

<u>Theorem 9.4.</u> If K is a compact Hausdorff space, then the following are equivalent:

(i) $C(K)$ is approximately finite dimensional.

(ii) K is totally disconnected and has a dense subset of isolated points.

(iii) If the extension $*M$ has the $\aleph_0$-isomorphism property, then $\widehat{C}(K)$ is isometric to $\hat{\ell}_\infty(n)$ for every infinite n in $*N$.

So for example c and $c \oplus_o c_0$ are $\equiv_A$ to ℓ_∞ and thus approximately finite dimensional, but $C[0,1]$ and $L^\infty[0,1]$ are not approximately finite dimensional. With the exception of $p = 2$, it is not known for any of the spaces $L^p[0,1]$, $1 \leq p < \infty$, whether it is approximately finite dimensional.

Next we turn to the examination of an interesting class of hyperfinite-dimensional Banach spaces, the spaces $\hat{\ell}_p(n)$ where both p and n are infinite. (If n is finite then $\hat{\ell}_p(n)$ is just $\ell_{\hat{p}}(n)$ where $\hat{p} = \mathrm{st}\, p$ with $\hat{p} = \infty$ if p is infinite. If n is infinite but p is finite, then $\hat{\ell}_p(n)$ may be naturally identified with $\hat{\ell}_{\hat{p}}(n)$, spaces which were discussed in Section 4.) In this case $\hat{\ell}_p(n)$ has a natural Banach lattice structure inherited from $\ell_p(n)$. Moreover if

$u,v \in \text{fin}(\ell_p(n))$ and have disjoint support, then

$$\text{st}\| u + v\| = \text{st}(\| u\|^p + \| v\|^p)^{1/p} = \max(\text{st}\| u\|, \text{st}\| v\|).$$

Hence $\hat{\ell}_p(n)$ is an abstract M-space. In the following we classify these spaces up to isometry (i.e., by $\equiv_A$) and discuss the standard implications of this classification.

We examine the order atoms in the space $\hat{\ell}_p(n)$ where p and n are infinite. (Recall that a is an order atom in a vector lattice L if and only if $x \in L$ and $|x| \leq |a|$ implies that x is a scalar multiple of a.) It is easy to see that an element a in $\hat{\ell}_p(n)$ is an atom of norm one exactly when it is of the form $\pm e_k$ for $1 \leq k \leq n$. These elements may be described in terms of the Banach space structure alone as exactly those elements a of $\hat{\ell}_p(n)$ such that there do not exist elements a_1 and a_2 in $\hat{\ell}_p(n)$ satisfying $\| a_1\| = 1$, $0 < \| a_2\| \leq 1$, $a = a_1 + a_2$ and $\| a_1 + (1/\| a_2\|)a_2\| = \| a_1 - (1/\| a_1\|)a_1\| = 1$.

Assume $n^{1/p}$ is finite with standard part α and let $u = \chi_{\{1,2,\ldots,n\}}$. Then $\pi(u)$ is a strong unit in $\hat{\ell}_p(n)$ (i.e., for each x in $\hat{\ell}_p(n)$ there exists β in R such that $|x| \leq \beta\pi(u)$) and $\| \pi(u)\| = \alpha$. If $\alpha = 1$, it is easy to see that $\hat{\ell}_p(n)$ may be identified with $\hat{\ell}_\infty(n)$. However, if $\alpha > 1$ then this is not the case; in fact as a Banach lattice $\hat{\ell}_p(n)$ fails to have the Fatou property since there exists an increasing net $u_\lambda \uparrow u$ such that $\sup_\lambda \| u_\lambda\| < \| u\|$. (Take as $\{u_\lambda\}$ the family of elements of the form χ_F, where F is a non-empty finite subset of $\{1,2,\ldots,n\}$. These elements all have norm one.)

Now let m be a positive infinite integer and q be a positive infinite element of $^*\mathbb{R}$ with $m^{1/q}$ finite but $\beta = \text{st}\, m^{1/q} \neq \alpha$. We assert that $\hat{\ell}_p(n)$ and $\hat{\ell}_q(m)$ are not isometric. Assume to the contrary that Φ is a linear isometry of $\hat{\ell}_p(n)$ onto $\hat{\ell}_q(m)$. Say $v \in \ell_q(m)$ such that $\Phi(\pi(u)) = \pi(v)$. Modifying Φ if necessary, we may assume $v \geq 0$ and $\| v\| = 1$. Then since the atoms of $\hat{\ell}_p(n)$ and $\hat{\ell}_q(m)$ may be described in terms of the Banach space structure alone, it follows that $\| \pi(v) + e_k\| = 2$ for $1 \leq k \leq m$. Hence $v(k) =_1 1$ for $1 \leq k \leq m$ and $\beta = \text{st}\| v\| = \|\pi(v)\|$ which is a contradiction.

On the other hand if $m^{1/q}$ is infinite then $\hat{\ell}_q(m)$ is not an atomic vector lattice. Indeed if $\alpha = (m^{1/q})^{-1}$ and $w = \alpha\,\chi_{\{1,2,\ldots,m\}}$ then $\pi(w)$ is a positive element of norm one which does not dominate any positive atoms. Thus for any norm-one atom e in $\hat{\ell}_q(m)$, we have $\| \pi(w) \pm e\| = 1$. It follows, also in this case, that $\hat{\ell}_q(m)$ is not isometric to $\hat{\ell}_p(n)$.

In fact the number $st(n^{1/p})$ (∞ if $n^{1/p}$ is infinite) separates the spaces $\hat{\ell}_p(n)$, n,p infinite, into equivalence classes under $\equiv_A$.

Theorem 9.5. Assume $*M$ has the $\aleph_0$-isomorphism property. Let p and q be infinite positive numbers in $*\mathbb{R}$ and let n and m be infinite positive integers in $*N$.

(i) If $n^{1/p}$ is finite, then $\hat{\ell}_p(n)$ is linearly isometric to $\hat{\ell}_q(m)$ if and only if $m^{1/q}$ is finite and $n^{1/p}$ and $m^{1/q}$ have the same standard part.

(ii) If $n^{1/p}$ is infinite then $\hat{\ell}_p(n)$ is linearly isometric to $\hat{\ell}_q(m)$ if and only if $m^{1/q}$ is also infinite.

Moreover, if $n^{1/p}$ is infinite, then $\hat{\ell}_p(n)$ is linearly isometric to $\hat{\ell}_q$ for all infinite q.

The necessity of the conditions (i) and (ii) was discussed above. The sufficency proof in (ii) is somewhat intricate and is given in [HMR 4]. We present here the proof that if $n^{1/p}$ and $m^{1/q}$ are both finite with the same standard part, then $\hat{\ell}_p(n)$ and $\hat{\ell}_q(m)$ are linearly isometric. The argument is similar to the one for part (ii) in its use of Theorem 8.8.

So assume $n^{1/p}$ and $m^{1/q}$ are both finite with the same standard part. Our task is to describe an extendable family of isometries of subspaces of $\hat{\ell}_p(n)$ onto $\hat{\ell}_q(m)$. We begin by identifying some simple finite dimensional subspaces of these spaces.

Definition 9.6. Let $\mathcal{B}$ be a finite Boolean algebra of internal subsets of $\{1,2,...,n\}$ (resp. $\{1,2,...,m\}$). The finite dimensional subspace S spanned by the characteristic functions of the elements of $\mathcal{B}$ is called the finite Boolean subspace of $\mathcal{B}$ and the atoms of $\mathcal{B}$ are called the generating sets for S.

Let G be the family of internal linear mappings T of a finite Boolean subspace dom(T) of $\ell_p(n)$ onto a finite Boolean subspace ran(T) of $\ell_q(m)$ such that A is a generating set for dom(T) if and only if $T(\chi_A) = \chi_B$ where B is a generating set for ran(T). Moreover if A is a generating set for dom(T) and $T(\chi_A) = \chi_B$ then (i) A is finite if and B is finite with the same cardinality and (ii) $|A|^{1/p} =_1 |B|^{1/q}$. It is easy to see that if $T \in G$, then T is an order isomorphism and $\|T\| =_1 \|T^{-1}\| =_1 1$. Define $\hat{T}$ mapping $(\mathrm{dom}(T))^\wedge$ onto $(\mathrm{ran}(T))^\wedge$ by setting $\hat{T}(\pi(p)) = \pi(T(p))$. We assert that the family $F = \{\hat{T}: T \in G\}$ is an extendable family of isometries of subspaces of $\hat{\ell}_p(n)$ into $\hat{\ell}_q(m)$.

Due to the symmetry of the definition of G, in order to establish the assertion it is sufficient to show that for each $T \in G$, for each p in $\mathrm{fin}(\mathrm{dom}(T))$ and each positive standard ε there exists an extension $\tilde{T}$ of T in G and $\tilde{p}$ in $\mathrm{dom}(\tilde{T})$ such that $\|p - \tilde{p}\| < \varepsilon$. We may assume $\|p\| < 1$. Let j be a positive finite integer such that $n^{1/p}/j < \varepsilon$ and for $-j \leq i \leq j-1$ define $C_i = \{k: 1 \leq k \leq n$ and $i/j \leq p(k) \leq (i+1)/j\}$. Let $\tilde{S}$ be the finite Boolean subspace having as generating sets, the non-empty sets of the form $A \cap C_i$, where A is a generating set for $\mathrm{dom}(T)$. If we define $\tilde{p}$ in $\tilde{S}$ by $\tilde{p} = \sum_{i=-j}^{j-1} (i/j)\chi_{C_i}$, then $\|p - \tilde{p}\| \leq \|1/j\, \chi_{\{1,2,\ldots,n\}}\| = (n^{1/p})/j < \varepsilon$.

It remains to show that T can be extended to $\tilde{T}$ on $\tilde{S}$ such that $\tilde{T} \in G$. For this it is sufficient to show that if A is a generating set for $\mathrm{dom}(T)$ with $T(\chi_A) = \chi_B$ and if $A = A_1 \cup A_2$ is a disjoint decomposition of A into internal sets then there is a disjoint decomposition $B = B_1 \cup B_2$ of B into internal sets such that (i) if either A_1 or A_2 is finite then the corresponding B_i is finite with the same cardinality and (ii) $|B_1|^{1/q} =_1 |A_1|^{1/p}$, $|B_2|^{1/q} =_1 |A_2|^{1/p}$. (A finite number of these steps yields $\tilde{T}$.) If either A is finite or A is infinite and A_1 (respectively A_2) is finite, the decomposition of B is immediate. So assume both A_1 and A_2 are infinite. If $|A_1|^{1/p} =_1 |A_2|^{1/p}$, construct B_1 and B_2 so that $|B_1|$ and $|B_2|$ differ by at most one. The final case to consider is $1 \leq \mathrm{st}|A_1|^{1/p} < \mathrm{st}|A_2|^{1/p} = \mathrm{st}|A|^{1/p}$. Let $B = \{b_1, b_2, \ldots b_\theta\}$ be an internal listing of B. Let ϕ be the smallest element i of $^*N$ such that $|\{b_1, b_2, \ldots, b_i\}|^{1/q} > \mathrm{st}(|A_1|^{1/p}) + 1/i$. Clearly i is infinite and if we set $B_1 = \{b_1, b_2, \ldots, b_\phi\}$ then $\mathrm{st}|B_1|^{1/q} = \mathrm{st}|A_1|^{1/p}$. Setting $B_2 = B \setminus B_1$, we are done.

Theorem 9.5, together with the results in [HMR 4] which treat the case $n^{1/p}$ infinite, have the following interpretation in terms of the standard $\ell_p(n)$ spaces and positive bounded sentences.

<u>Theorem 9.7.</u> For a positive bounded sentence σ and an integer k in N there exists an integer $M = M(\sigma, k)$ in N such that if $m, n, p, q \geq M$ and $|m^{-1/p} - n^{-1/q}| \leq 1/M$ then σ_k is true in $\ell_p(n)$ whenever σ is true in $\ell_q(m)$.

For another interpretation of these results in purely standard terms, see Section 8 (Theorem 8.3).

These results give typical examples of how results comparing standard Banach spaces can be simply derived from classification results for nonstandard hulls. See also [HEN 3, Theorem 2.5 and Theorem 3.8].

Section 10. Krivine's Theorem.

In this Section we give a modified version of the proof by H. Lemberg [LE] of a deep result first proved by J.-L. Krivine [KR 3]. This result gives the best information so far about the way in which the fundamental spaces c_0 and $\ell_p (1 < p < \infty)$ occur in a completely general infinite dimensional Banach space. The proofs by Krivine and Lemberg are phrased in terms of ultraproducts of Banach spaces. Another proof of Krivine's theorem was given by H. P. Rosenthal [ROS 2].

Theorem 10.1. (Krivine) Let E be a standard Banach space and (x_n) a sequence in E whose linear span is infinite dimensional. Then the usual basis of c_0 or of ℓ_p (for some $1 \leq p < \infty$) is block-finitely represented in (x_n).

Remark. A sequence (y_n) is block finitely represented in (x_n) if for each $m \in \mathbb{N}$ and each $\varepsilon > 0$ there exist integers $1 = k_1 < k_2 < \dots < k_{m+1}$ and vectors $z_1, \dots, z_m$ so that z_j is equal to a linear combination of the x_j for which $k_j \leq j < k_{j+1}$ and so that the sequences $(y_1, \dots, y_m)$ and $(z_1, \dots, z_m)$ are $(1+\varepsilon)$-equivalent. (Note that this relation is clearly transitive.)

Given the smooth way in which the concept of finite representability is handled using nonstandard hulls (see Section 3), it is not surprising that the concept defined here can be handled by the same methods.

We begin the proof of Theorem 10.1 by making some reductions. First of all, we may assume that (x_n) is bounded in norm and that $\| x_n - x_m \| \geq \delta > 0$ whenever $n \neq m$, for some δ. (That is, we can find a sequence in the span of the original sequence which satisfies these conditions and is block finitely represented in the original sequence.) This is trivial since the span of the original sequence is infinite dimensional.

Second, we may assume that the sequence (x_n) is subsymmetric. (That is, for any sequence of coefficients $\alpha_1, \dots, \alpha_m$ and any indices $i_1 < \dots < i_m$

$$\| \sum_{j=1}^{m} \alpha_j x_j \| = \| \sum_{j=1}^{m} \alpha_j x_{i_j} \| .)$$

This is done using the techniques of Brunel and Sucheston [BRS 1] based on Ramsey's Theorem. They show that our sequence (x_n) possesses a subsymmetric sequence (y_n) in the sense that (y_n) is even finitely represented in (x_n). (That is, in the definition we can always take

(z_j) to be a subsequence of (x_n).) Finally we can assume that the given sequence (x_n) converges weakly to 0. Here we use a celebrated theorem of Rosenthal [ROS 1]. This result asserts that either (x_n) has a subsequence that is asymptotically equivalent to the ℓ_1 basis (in which case we are done) or it has a weakly Cauchy subsequence (x_n'). In that case we replace (x_n) by the sequence of differences $x'_{2n+1} - x'_{2n}$.

The end result of these simple reductions is that in proving Theorem 10.1 we may assume (x_n) is subsymmetric, weakly null, that $\|x_n\| = 1$ for all n and that $\|x_n - x_m\| \geq \delta > 0$ whenever $n \neq m$. It follows from this that (x_n) is also <u>1-unconditional</u> [BRS 2]. (That is, for each sequence of coefficients $\alpha_1,\ldots,\alpha_m$ and each set $F \subseteq \{1,2,\ldots,m\}$

$$\|\sum_{j=1}^{m} \alpha_j x_j\| \geq \|\sum_{j\in F} \alpha_j x_j\| .)$$

In particular, for each sequence of coefficients $\alpha_1,\ldots,\alpha_m$ we have

$$\|\sum_{j=1}^{m} \alpha_j x_j\| = \|\sum_{j=1}^{m} |\alpha_j| x_j\| .$$

As discussed in Section 1, we are thinking of the given space E as a real Banach space and all the conditions discussed above are considered for real scalars only. However, Lemberg's ideas depend on the introduction of complex coefficients. Elementary operator theory is applied to certain operators on the complex linear span of the sequence (x_n).

Let F denote a complexification of the linear span of (x_n); that is, F consists of all complex linear combinations $\Sigma\, \alpha_i x_i$ for which the sum $\Sigma |\alpha_i| x_i$ is in E. The norm on F is defined by

$$\|\Sigma\, \alpha_i x_i\| = \|\Sigma |\alpha_i| x_i\| .$$

Next we pass from the subsymmetric sequence (x_n) to a system $(e_r | r$ is a real number $0 < r < 1)$ with a norm defined on complex linear combinations by

$$\|\sum_{j=1}^{m} \alpha_j e_{r_j}\| = \|\sum_{j=1}^{m} \alpha_j x_j\|$$

as long as $r_1 < r_2 < \ldots < r_m$, which we can certainly achieve by rearranging terms. Evidently the complex linear span of $(e_r | 0 < r < 1)$ with this norm gives rise by completion to a Banach space, which we will denote by X. If $r_1 < r_2 < \ldots$ is a sequence of real numbers between 0 and 1, then the sequence (e_{r_j}) is isometrically equivalent to the sequence (x_n). Hence it suffices to prove that c_0 or some ℓ_p $(1 \leq p < \infty)$ is block finitely represented in $(e_r | 0 < r < 1)$, using the transitivity of this relation.

All this is quite standard. The innovation due to Lemberg is to introduce certain operators on X. For each integer k we define T_k on each e_r by

$$T_k(e_r) = e_{r_1} + \ldots + e_{r_k}$$

where $r_j = \frac{r+j-1}{k}$ for each $j = 1,2,\ldots,k$. Note that $0 < r_1 < r_2 < \ldots < r_k < 1$ so that $\| T_k(e_r)\| = \| x_1 + \ldots + x_k \|$ and does not depend on r. (We have modified Lemberg's argument slightly by introducing T_k for every k, instead of just T_2 and T_3. Also we make the role of the complexification more explicit.) Now we extend T_k linearly to be defined on the span in X of $(e_r | 0 < r < 1)$. We observe that for any $x = \sum_{j=1}^{m} \alpha_j e_{r_j}$, T_k satisfies the inequalities

$$\| x \| \leq \| T_k(x) \| \leq k \cdot \| x \| .$$

The left inequality follows from the 1-unconditionality of $(e_r | 0 < r < 1)$ (inherited from (x_n)) and the subsymmetry. The right inequality follows from the subsymmetry and the triangle inequality. For example, say $x = \alpha e_r + \beta e_s$, $r < s$. Then

$$T_k(x) = \alpha e_{r_1} + \alpha e_{r_2} + \ldots + \alpha e_{r_k} + \beta e_{s_1} + \beta e_{s_2} + \ldots + \beta e_{s_k}$$

where $r_j = (r+j-1)/k$ and $s_j = (s+j-1)/k$. Note that $r_j < s_j$ holds for each j. We write

$$T_k(x) = \Sigma(\alpha e_{r_j} + \beta e_{s_j})$$

which is the sum of k terms, each of which has the same norm as $\alpha e_r + \beta e_s$, by subsymmetry.

From these inequalities we see that T_k extends to a bounded linear operator on all of X, which satisfies the same norm inequalities. Moreover, an easy calculation shows that the T_k's form a commuting system of operators. Indeed, for any $k, \ell \in \mathbb{N}$ we see that

$$T_k \circ T_\ell = T_{k\ell} = T_\ell \circ T_k$$

on X.

We summarize the properties of $\{T_k \mid k \in \mathbb{N}\}$ in Lemma 10.2 below. It will be useful to introduce two concepts first: let $x = \sum_{j=1}^{n} \alpha_n e_{r_j}$ and $y = \sum_{i=1}^{m} \beta_i e_{s_i}$ be finite linear combinations of $(e_r \mid 0 < r < 1)$ with complex coefficients. We write

$$\operatorname{supp}(x) < \operatorname{supp}(y)$$

if $r_j < s_i$ holds for every j and i. Also we say that x and y are <u>congruent</u> if $r_1 < r_2 < \ldots < r_n$, $s_1 < s_2 < \ldots < s_m$, $m = n$ and $\alpha_i = \beta_i$ for each i. In that case $\|x\| = \|y\|$ by subsymmetry. Also if

$$\operatorname{supp}(x) < \operatorname{supp}(z_1) < \operatorname{supp}(y),$$
$$\operatorname{supp}(x) < \operatorname{supp}(z_2) < \operatorname{supp}(y)$$

and if z_1, z_2 are congruent, then also $x + z_1 + y$ and $x + z_2 + y$ are congruent (so

$$\|x + z_1 + y\| = \|x + z_2 + y\|$$

in that case).

<u>Lemma 10.2.</u> $\{T_k \mid k \in \mathbb{N}\}$ is a commuting family of bounded linear operators on X. In particular

(1) $T_k \circ T_\ell = T_{k\ell} = T_\ell \circ T_k$ for each $k, \ell \in \mathbb{N}$, and

(2) $\|x\| \le \|T_k(x)\| \le k\|x\|$ for each $x \in X$ and $k \in \mathbb{N}$.

Moreover, for each $x = \sum_{j=1}^{n} \alpha_j e_{r_j}$ and each $k \in \mathbb{N}$,

$$T_k(x) = x_1 + x_2 + \ldots + x_k$$

where $x_1, x_2, \ldots, x_k$ are pairwise congruent and $\operatorname{supp}(x_1) < \operatorname{supp}(x_2) < \ldots < \operatorname{supp}(x_k)$.

We now pass to the nonstandard hull $\hat{X}$ and to the linear operators $\hat{T}_k$ on $\hat{X}$. Recall that they are defined very simply by

$$\hat{T}_k(x) = \pi({}^*T_k(p))$$

where $x \in \hat{X}$ satisfies $x = \pi(p)$. It is immediate that $\{\hat{T}_k\}$ satisfies (1) and (2) in Lemma 10.2. Moreover, they also satisfy a version of the last clause of Lemma 10.2, where $\sum_{j=1}^{n} \alpha_j e_{r_j}$ is taken to be a

* finite sum, $\alpha_1,\ldots,\alpha_n$ is an internal sequence from $^*\mathbb{C}$ and $r_1,\ldots, r_n$ is an internal sequence of numbers in the nonstandard interval $^*(0,1)$. Note that the concepts "supp( ) < supp( )" and "congruent" have an obvious transferred meaning applied to such *-finite sums. Moreover, for each $x \in \hat{X}$ there exists such a *-finite sum $\sum_{j=1}^{n} \alpha_j e_{r_j}$ such that

$$x = \pi(\Sigma\, \alpha_j\, e_{r_j}) ,$$

since the finite linear combinations of $(e_r | 0 < r < 1)$ are dense in X.

Lemma 10.3. There exists a common eigenvector for the operators $\{\hat{T}_k | k \in \mathbb{N}\}$ on $\hat{X}$.

Proof. (This corresponds to the key innovation due to Lemberg [LE] and amounts to a very interesting amalgamation of combinatorics and functional analysis.)

We use very elementary facts from operator theory (see for example [DO, pages 3-9]) together with a trivial saturation argument. First choose λ_2 to be an approximate eigenvalue of $\hat{T}_2$ (for example, a boundary point of the spectrum of $\hat{T}_2$). This means that there exists a sequence (z_m) of norm 1 vectors in $\hat{X}$ such that $\|\hat{T}_2(z_m)-\lambda_2 z_m\| \to 0$ as $m \to \infty$. Hence by a saturation argument $\hat{T}_2$ actually has an eigenvector with eigenvalue λ_2. Let $Y_2 \subseteq \hat{X}$ be the subspace of all such eigenvectors

$$Y_2 = \{x \in \hat{X} | \hat{T}_2(x) = \lambda_2 x\} .$$

Because the operators $\{\hat{T}_k\}$ commute, each $\hat{T}_k$ leaves Y_2 invariant. We next let λ_3 be an approximate eigenvalue of $\hat{T}_3$ restricted to Y_2. A similar argument shows that $\hat{T}_3$ has an eigenvector in Y_2 and this is a common eigenvector for $\hat{T}_2$ and $\hat{T}_3$. Proceeding inductively and applying one final saturation argument at the end yields the common eigenvector for the entire system $\{\hat{T}_k | k \in \mathbb{N}\}$.

Notation. Let $(\lambda_n | n \in \mathbb{N})$ be a sequence of complex numbers chosen as in the argument above and let $z \in \hat{X}$ be a common eigenvector of $\{\hat{T}_k\}$; that is $T_k(z) = \lambda_k z$ all $k \in \mathbb{N}$. We may assume $\|z\| = 1$. Moreover, let $u = \sum_{j=1}^{m} \alpha_j e_{r_j}$ be a *-finite sum in $^*X$ so that $\pi(u) = z$. Finally let $u_1,u_2,u_3,\ldots$ be a sequence of *-finite sums in $^*X$ such that (i) each u_j is congruent to u; (ii) $\text{supp}(u_i) < \text{supp}(u_j)$ for all $i < j$. Let $z_j = \pi(u_j)$ for each $j \in \mathbb{N}$.

A straightforward transfer argument shows that the sequence (z_n) is block finitely represented in $(e_r | 0 < r < 1) \subseteq X$, and hence in the original sequence (x_n). Therefore it will complete our proof of Theorem 10.1 to prove the following:

<u>Lemma 10.4.</u> Let (z_n) be as above.

(1) If $|\lambda_2| = 1$ then the real linear span of (z_n) is isometrically equivalent to the usual basis of c_0.

(2) If $|\lambda_2| = 2^{1/p}$, $1 \leq p < \infty$, then the real linear span of (z_n) is isometrically equivalent to the usual basis of ℓ_p.

<u>Remark.</u> We chose λ_2 to be an arbitrary approximate eigenvalue of $\hat{T}_2$. It is easily seen that these are exactly the same as the approximate eigenvalues of T on X. Hence some values of p for which the usual basis of ℓ_p is block finitely represented in (x_n) can be calculated directly from the spectrum of T_2. This may be useful since T_2 is a relatively simple operator to study. Also it raises the question whether all the values of p for which the ℓ_p basis is block finitely represented in (x_n) arise in this way from the approximate eigenvalues of T_2.

<u>Remark.</u> The eigenvalues $\{\lambda_k | k \in \mathbb{N}\}$ satisfy the multiplicative identity

$$\lambda_k \cdot \lambda_\ell = \lambda_{k\ell}$$

for all $k, \ell \in \mathbb{N}$. (Indeed,

$$\hat{T}_k(\hat{T}_\ell(z)) = \lambda_k \lambda_\ell \cdot z = \hat{T}_{k\ell}(z) = \lambda_{k\ell} \cdot z.)$$

Moreover, $1 = |\lambda_1| \leq |\lambda_2| \leq |\lambda_3| \leq \dots$. To see this we need only note that if u is as above (so that $\pi(u) = z$) then for each $k \in \mathbb{N}$, ${}^*T_k(u)$ is the sum of k vectors, each congruent to u and with their supports properly ordered (as in Lemma 10.2). Therefore, by the 1-unconditionality of $(e_r | 0 < r < 1)$, it follows that

$$\| {}^*T_k(u) \| \leq \| {}^*T_\ell(u) \|$$

whenever $k \leq \ell$. Passing to $\hat{X}$ we obtain

$$|\lambda_k| \cdot \| z \| = \| \hat{T}_k(z) \| \leq \| \hat{T}_\ell(z) \| = |\lambda_\ell| \cdot \| z \|$$

so that $|\lambda_k| \leq |\lambda_\ell|$, as desired.

It is now an elementary exercise to show that the real numbers

$\{|\lambda_k|\}$ must satisfy the identity

$$|\lambda_k| = k^\alpha \quad \text{all} \quad k \in \mathbb{N}$$

where $\alpha \geq 0$ is chosen so that $|\lambda_2| = 2^\alpha$. Note $\alpha \leq 1$ since $\|T_2\| \leq 2$.

Proof of Lemma 10.4. Let $\alpha \geq 0$ be as above, so that the eigenvalues λ_k satisfy $|\lambda_k| = k^\alpha$ for all $k \in \mathbb{N}$. If $\alpha = 0$ then every $|\lambda_k| = 1$; this corresponds to case (1) in the Lemma. If $\alpha > 0$ then we set $p = 1/\alpha$; this corresponds to case (2).

(We will treat the $\alpha > 0$ case in detail; the other case uses the same ideas. Fix a standard integer n and standard positive coefficients $r_1,\ldots,r_n$. Let M be a common denominator for these numbers and write $r_j = k_j/M$, so $k_1,\ldots,k_n \in \mathbb{N}$. We will estimate the norm of the vector

$$w = k_1^\alpha u_1 + k_2^\alpha u_2 + \ldots + k_n^\alpha u_n .$$

Fix j $(1 \leq j \leq n)$ and collect terms in this vector so it is written in the form

$$w = x + k_j^\alpha \cdot u_j + y.$$

Note that

$$\operatorname{supp}(x) < \operatorname{supp}(u_j) < \operatorname{supp}(y).$$

We may change basis vectors in w without changing coefficients, so as to obtain a vector

$$w' = x' + k_j^\alpha \cdot u + y'$$

which is congruent to w. We require that x' be congruent to x, y' to y and that

$$\operatorname{supp}(x') < \operatorname{supp}(u) < \operatorname{supp}(y')$$

and that

$$\operatorname{supp}(x') < \operatorname{supp}(u_i) < \operatorname{supp}(y')$$

for all $1 \leq i \leq k_j$. Now $k_j^\alpha \cdot u$ is infinitely close in norm to ${}^*T_{k_j}(u)$, since $|\lambda_j| = k_j^\alpha$ and $\hat{T}_{k_j}(z) = \lambda_j \cdot z$. Moreover ${}^*T_{k_j}(u)$ is congruent to

$$u_1 + u_2 + \ldots + u_{k_j}$$

by Lemma 10.2. Therefore $\|w\| = \|w'\|$ is infinitely close to

$$\|x' + u_1 + \ldots + u_{k_j} + y'\| .$$

(Here we are using the fact that we defined our complexification in such a way that $\| \Sigma\, \alpha_i\, e_{r_i} \| = \| \Sigma\, \beta_i\, e_{r_i} \|$ whenever $\{\alpha_i\}$ and $\{\beta_i\}$ are complex coefficients satisfying $|\alpha_i| = |\beta_i|$ for all i.) If we carry out this "expansion" process for each $j = 1,2,\ldots,n$ inductively, we find that our original vector w has norm infinitely close to

$$\| u_1 + u_2 + \ldots + u_N \|$$

where $N = k_1 + k_2 + \ldots + k_n$. This norm is infinitely close to $N^\alpha = |\lambda_N|$, since $u_1 + \ldots + u_N$ is congruent to ${}^*T_N(u) \approx \lambda_N \cdot u$. Thus we have shown that

$$\| \sum_{j=1}^{n} k_j^\alpha \cdot u_j \| \approx \left(\sum_{j=1}^{n} k_j \right)^\alpha .$$

If we now divide by M^α, write s_j for $r_j^\alpha = (k_j/M)^\alpha$ and recall $p = 1/\alpha$, this yields

$$\| \sum_{j=1}^{n} s_j \cdot u_j \| \approx \left(\sum_{j=1}^{n} s_j^p \right)^{1/p}$$

and hence, passing to $\hat{X}$, we get

$$\| \Sigma\, s_j\, z_j \| = (\Sigma\, s_j^p)^{1/p} .$$

Since numbers of the form r^α (r positive rational) are dense in the positive real line ($\alpha > 0$), it follows that this norm equality holds for all positive real numbers $s_1,\ldots,s_n$. Since (z_n) is 1-unconditional (over real scalars) the proof of Lemma 10.2(2) is complete.

Note. Basic ideas for the norm estimates used above go back to the fundamental paper of Zippin [ZI], where one can also find details of the c_0 case.

Section 11. Lipschitz and Uniform Equivalence of Banach Spaces.

In this Section we briefly indicate how S. Heinrich and P. Mankiewicz [HMK] have used methods of nonstandard analysis to treat nonlinear equivalence between Banach spaces. (Formally their work is phrased in the language of ultraproducts.) We will particularly concentrate on questions of uniform equivalence. First we prove an elementary result showing how passage to nonstandard hulls converts uniform equivalence into Lipschitz equivalence, which can then be handled using familiar methods of standard analysis. This idea is a simple application of the observation due to Corson and Klee [CK] that uniformly continuous mappings on metric spaces obey Lipschitz estimates for arguments that are far apart.

<u>Lemma 11.1.</u> Let V and W be internal Banach spaces in $*M$ and suppose the nonstandard hulls $\hat{V}$, $\hat{W}$ are uniformly homeomorphic. Then $\hat{V}$, $\hat{W}$ are Lipschitz homeomorphic.

Therefore, if E, F are uniformly homeomorphic standard Banach spaces, then their nonstandard hulls $\hat{E}$, $\hat{F}$ are Lipschitz homeomorphic.

<u>Proof.</u> We assume $*M$ has the $\aleph_0$-isomorphism property. Let G be a uniform homeomorphism from $\hat{V}$ onto $\hat{W}$. Using a downward Löwenheim-Skolem theorem, we may obtain standard Banach spaces E, F and a uniform homeomorphism g from E onto F, with $E \equiv_A \hat{V}$ and $F \equiv_A \hat{W}$. (Compare Theorem 5.6.) Then $\hat{E}$ is linearly isometric to $\hat{V}$, $\hat{F}$ is linearly isometric to $\hat{W}$ and $\widehat{*g}$ is a uniform homeomorphism from $\hat{E}$ onto $\hat{F}$. Since it evidently suffices to show here that $\hat{E}$ and $\hat{F}$ are Lipschitz homeomorphic, we will work with $\widehat{*g}$ instead of the original G. (This permits us to use nonstandard analysis methods on the uniform isomorphism, where this may not have been possible to do on the original G. This illustrates another way in which the $\aleph_0$-isomorphism property can be technically useful.)

Since g is a standard uniformly continuous mapping, there exists a standard $1 > \delta > 0$ so that for $p,q \in *E$

$$\|p - q\| \le \delta \Rightarrow \|*g(p) - *g(q)\| \le 1.$$

(A similar fact is true of g^{-1} also.) Now fix $p,q \in *E$ with $\|p - q\| \ge 1$. Choose $n \in *\mathbb{N}$ (n need not be standard) so that

$$\frac{\|p - q\|}{\delta} \le n < \frac{2\|p - q\|}{\delta} .$$

Then let $p_0, p_1, \ldots, p_n$ be determined by setting $p_0 = p$ and $p_{j+1} = p_j + 1/n(q - p)$ for $j \ge 0$. Then $\|p_{j+1} - p_j\| \le \delta$ for each j, so that by the triangle inequality,

$$\|*g(p) - *g(q)\| \le n < (\tfrac{2}{\delta})\|p - q\| .$$

Let $K = 2/\delta$, a standard number. (Again we do the same analysis for g^{-1}.)

Now we define a mapping f from $*E$ to $*F$ by

$$f(p) = \frac{1}{H} \cdot *g(H_p)$$

where H is a fixed, infinite integer > 0. (Note that f maps onto

$*F$ since $*g$ does and $f^{-1}(q) = 1/H \, *g^{-1}(Hq)$, so our situation continues to be symmetric.)

We will show that $\hat{f}$ is a Lipschitz homeomorphism from $\hat{E}$ onto $\hat{F}$ (with Lipschitz constant K.) Because of the symmetry between f and f^{-1}, it suffices to show

(a) if p,q are finite and $\|p - q\| \not\approx 0$, then $\|f(p) - f(q)\| \leq K \|p - q\|$, and

(b) if p,q are finite and $\|p - q\| \approx 0$, then $\|f(p) - f(q)\| \approx 0$.

(Note that (a) implies that f maps finite elements of $*E$ to finite elements of $*F$ and (b) shows that $\hat{f}$ is well-defined.)

Proof of (a): If $\|p - q\|$ is not infinitesimal, then it is $\geq 1/H$. Therefore $\|Hp - Hq\| \geq 1$. From the calculation above we see that

$$\|f(p) - f(q)\| = (1/H)\|*g(Hp) - *g(Hq)\|$$
$$\leq (K/H)\|Hp - Hq\| = K\|p - q\|$$

Proof of (b): Since g is uniformly continuous we know that $\|p-q\| \approx 0$ implies $\|*g(p) - *g(q)\| \approx 0$. Given finite elements p,q of $*E$ with $\|p - q\| \approx 0$, define an internal sequence $p_0, p_1, \cdots, p_H$ with $p_0 = Hp$, $p_{j+1} = p_j + (q - p)$ for all j, so that $p_H = H_q$. Then we calculate

$$\|f(p) - f(q)\| = (1/H)\|*g(Hp) - *g(Hq)\|$$
$$\leq (1/H) \sum_{j=0}^{H-1} \|*g(p_j) - *g(p_{j+1})\|$$
$$\leq (1/H) \cdot H \cdot \varepsilon = \varepsilon$$

where ε is the maximum of the norms $\|*g(p_j) - *g(p_{j+1})\|$ for $0 \leq j < H$. These norms are all ≈ 0 (by the uniform continuity of g) so that $\varepsilon \approx 0$ and the proof is complete.

Next we show how Heinrich and Mankiewicz use this approach to give a new proof of a theorem of Ribe [RI 1,2]. This result and others show how closely the linear structure of a Banach space is tied to the nonlinear uniform structure.

Theorem 11.2. If E and F are uniformly homeomorphic standard Banach spaces, then each of E, F is λ-finitely represented in the other, for some $\lambda \geq 1$.

<u>Proof.</u> By symmetry it suffices to show that E is λ-finitely represented in F for some $\lambda > 1$. Let $X \subseteq E$ be any finite dimensional subspace. We will sketch the proof that X can be λ-isomorphically embedded in the second dual space $(\hat{F})''$ with λ independent of X. By Proposition 3.8 and the Principle of Local Reflexivity we obtain that E is λ-finitely represented in F.

By Lemma 11.1 we can assume that there exists a Lipschitz homeomorphism from $\hat{E}$ onto $\hat{F}$. In particular, since $X \subseteq E \subseteq \hat{E}$, there is a Lipschitz homeomorphism G from X onto a subset of $\hat{F}$.

Using the Downward Löwenheim-Skolem Theorem (Theorem 5.6), find a separable space $Z \subseteq (\hat{F})'$ such that Z is norming for the set $G(X) \subseteq \hat{F}$ and Z is a reflecting subspace of $(\hat{F})'$. By Corollary 3.15 the dual space Z' can be embedded into $(\hat{F})''$ by a linear isometry. Also, since Z was chosen to norm $G(X)$, there is an isometric mapping from $G(X)$ into Z'.

That is, we have a separable space Z whose dual Z' is linearly isometric to a subspace of $(\hat{F})''$ and a Lipschitz homeomorphism (which we still call G) from X onto a subset of Z'. Let $\{z_n\}$ be a dense subset of Z. For each n let f_n be defined on X by

$$f_n(x) = \langle z_n, G(x) \rangle$$

for $x \in X$. The mappings $\{f_n\}$ are all Lipschitz mappings (with the same Lipschitz constant as the original G) from the finite dimensional space X into $\mathbb{R}$.

We now apply a classical theorem due to Rademacher: a Lipschitz mapping from $\mathbb{R}^n$ to $\mathbb{R}$ is differentiable at almost every point of $\mathbb{R}^n$. This implies that there is a single point x_0 in X at which all of the mappings $\{f_n\}$ are differentiable. From the system of differentials $\{D_x f_n\}$ it is possible to construct a linear isomorphism from X into Z' (and hence into $(\hat{F})''$). For the details of how this is done, see [HMK].

<u>Remark.</u> In this argument, the elementary calculations based on nonstandard analysis lead to a situation where one can apply a classical theorem of real analysis. In the original proof due to Ribe one finds very detailed combinatorial arguments. This is characteristic of the ways in which Heinrich and Mankiewicz have used these methods in their general study of nonlinear problems.

In the remainder of this Section we indicate a few of the many new results obtained by S. Heinrich and P. Mankiewicz in [HMK]. Their study of uniformly homeomorphic Banach spaces uses Theorem 10.1 to permit concentration on Lipschitz homeomorphisms together with a close analysis of Lipschitz mappings and their differentials.

One of their fundamental technical results is the following:

Theorem 11.3. Suppose E, F are standard Banach spaces and one of them is super-reflexive. Suppose also that E, F satisfy one of the hypotheses of the Pelczynski Decomposition Scheme. (For example either (a) E and F are linearly isomorphic to their squares or (b) one of them, say E, has a complemented subspace linearly isomorphic to $\ell_p(E)$ for some $1 \leq p < \infty$.)

Under these hypotheses, if E and F are uniformly homeomorphic, their nonstandard hulls $\hat{E}$ and $\hat{F}$ are linearly isomorphic.

Remark. The Pelczynski Decomposition Scheme enters in the following way. If E, F are uniformly homeomorphic, then the methods discussed above are used to linearly embed E as a complemented subspace of $\hat{F}$ and F as a complemented subspace of $\hat{E}$. It follows easily (when *M satisfies the $\aleph_0$-isomorphism property) that $\hat{E}$, $\hat{F}$ can each be linearly embedded as a complemented subspace of the other. The Pelczynski Decomposition Method is then used to show that $\hat{E}$, $\hat{F}$ are linearly isomorphic to each other. Note that it is only necessary to assume that $\hat{E}$, $\hat{F}$ satisfy some hypothesis of the Decomposition Scheme. (See [LT 1] for discussion of this method of proof.)

Corollary 11.4. If E is a super-reflexive rearrangement invariant function space on [0,1] and if E, F are uniformly homeomorphic, then the nonstandard hulls $\hat{E}$, $\hat{F}$ are linearly isomorphic.

Corollary 11.5. Suppose $1 < p < \infty$ and E is a $\mathcal{L}_p$-space. If E is uniformly homeomorphic to F, then $\hat{E}$ and $\hat{F}$ are linearly isomorphic. In particular, F is also an $\mathcal{L}_p$-space.

Proof. By Theorem 6.6(a) $\hat{E}$ is linearly isomorphic to $\hat{\ell}_p$ where E is an $\mathcal{L}_p$-space and $1 < p < \infty$. Also the pair $\hat{F}$, $\hat{\ell}_p$ satisfy a hypothesis of the Pelczyinski Decomposition Scheme, since $\hat{\ell}_p$ is $L_p(\mu)$ for some μ. Hence Theorem 10.4 applies.

Corollary 11.6. If E is a $\mathcal{L}_\infty$-space and E is uniformly homeomorphic to F, then F is also a $\mathcal{L}_\infty$-space.

Section 12. Problems.

Problem 1. Characterize the Banach spaces E such that $E \equiv_A C([0,1])$. By the result of S. Heinrich [HEI 5] E must be isometric to a $C(K)$ space. By [HEN 3, Corollary 3.4] K must be connected.

This type of question is open for most other compact spaces. Only for totally disconnected spaces K is there a characterization of the spaces E for which $E \equiv_A C(K)$. (See Theorem 6.2.)

Problem 2. Give explicitly a set Σ of positive bounded sentences with the property that E is (isometric to) a $C(K)$-space if and only if $E \models_A \Sigma$.

Problem 3. Prove a version of the local duality theorem of Kürsten and Stern for Banach lattices (See Theorem 7.1 and Corollary 7.2). (Added in proof: this has been done by K. D. Kürsten.)

Problem 4. Characterize the Banach spaces E such that some nonstandard hull of E has Banach lattice structure? (respectively, has lattice structure under which it is an abstract M-space?) In [HHM] a Banach space E_0 is constructed such that E_0 does not have Banach lattice structure while the nonstandard hull $\hat{E}_0$ is isometric to $\hat{c}_0$ and thus has abstract M-space structure.

The space E_0 is nonseparable. Is it true that if E is separable and $\hat{E}$ has Banach lattice structure (resp., abstract M-space structure) then E also has the same structure?

Problem 5. Does there exist a *-finite dimensional internal Banach space V such that $\hat{V}$ is a (nonseparable) Gurarii space. Equivalently, does there exist such a V with $\hat{V}$ linearly isometric to $\hat{E}$, where E is the separable Gurarii space? (See Section 6.)

A standard problem equivalent to this one is the following: given an arbitrary integer k and an arbitrary $\lambda > 1$, does there exist a finite dimensional space $E = E(k,\lambda)$ with the property that whenever $X \subseteq E$ and Y is a Banach space of dimension $\leq k$ with $X \subseteq Y$, then there is a λ-isomorphic embedding $T: Y \to E$ which is the identity on X?

Problem 6. Does there exist a *-finite dimensional internal Banach space V such that $\hat{V}$ is linearly isometric to $\hat{L}_1$; that is, is $\hat{L}_1$ linearly isometric to a hyperfinite dimensional space? This problem is also open for $\hat{L}_p$ except when $p = 2$ or ∞.

This can be restated equivalently in various ways. (See [HEN 3] and Section 9.) Note that this is a purely isometric question: by Theorem 6.6 the spaces $\hat{L}_p$ and $\hat{\ell}_p$ are isomorphic (though not isometric) so that $\hat{L}_p$ is isomorphic to $\hat{\ell}_p(n)$ for any infinite $n \in {}^*\mathbb{N}$.

Problem 7. Is it true that for every Banach space E, there is a $*$-finite dimensional internal Banach space V such that $\hat{V}$ is linearly isomorphic to $\hat{E}$; that is, is every nonstandard hull isomorphic to a hyperfinite dimensional space?

If so, this would tend to show that the isomorphic local theory of Banach spaces is a purely finite dimensional subject. In [MO 2], [MO 3] it is shown that $\hat{c}_0$, $\hat{C}[0,1]$ and many other nonstandard hulls are not isometric to hyperfinite dimensional spaces. However, they (and almost all specific examples that have been studied) are known to be isomorphic to such spaces.

Problem 8. Let E be a Banach space with an unconditional basis (x_n) and let E_n be the linear span of $x_1, x_2, \ldots, x_n$. Suppose that E is approximately finite dimensional (i.e., there is a $*$-finite dimensional space V such that $\hat{V}$ and $\hat{E}$ are isometric). Must there exist an infinite $\omega \in {}^*\mathbb{N}$ such that $\hat{E}$ is isometric in fact to ${}^*\hat{E}_\omega$?

In the case where E is ℓ_p for some $1 \leq p \leq \infty$, it is known that ℓ_p is approximately finite dimensional and for these spaces the answer to this problem is positive, [HEN 1]. When E is c_0, then E is not approximately finite dimensional and the differences in structure between $\hat{E}$ and ${}^*\hat{E}_\omega$ in this case are an important part of the proof.

Problem 9. (S. Heinrich) If a nonstandard hull $\hat{E}$ has Grothendieck's Approximation Property, must E (or equivalently $\hat{E}$) have the Uniform Approximation Property?

See [HEI 4, Section 9] for a discussion of this problem.

Problem 10. Suppose V is any internal Banach space. For which compact linear operators T on $\hat{V}$ does there exist an internal linear operator S on V such that $\hat{S}$ and T have the same range, the same norm and agree on some given separable subspace of $\hat{V}$? (See Section 7.)

Problem 11. Suppose V, W are internal Banach spaces and $T: V \to W$ is an internal linear operator such that $\|T\|$ is finite and the range of T has standard finite dimension. Is it true that

$$n(\hat{T}) = st(n(T))?$$

(See Section 7.)

If the dimension of the range of T is m, then the proof of Lemma 7.3 shows that

$$\|T\| \leq n(T) \leq m\|T\|.$$

(Here these concepts are being considered internally in $*M$. Hence $n(T)$ is the inf in $*\mathbb{R}$ of all *-finite sums

$$\sum_{j=1}^{\omega} \|\phi_j\| \cdot \|f_j\|$$

where $\phi_1,\phi_2,\ldots,\phi_\omega$ and $f_1,f_2,\ldots,f_\omega$ are internal sequences, from V' and W resp., and T is represented by

$$T(x) = \sum_{j=1}^{\omega} \phi_j(x) \cdot f_j$$

for all $x \in V$. Since $\|T\|$ is assumed to be finite, $\hat{T}: \hat{V} \to \hat{W}$ is well defined with $\|\hat{T}\| = st\|T\|$. Also it is easy to show that the range of $\hat{T}$ has dimension m, so $n(\hat{T})$ is defined.

Problem 12. Given internal Banach spaces V and W, find representations for the nonstandard hulls of the internal spaces of operators $L(V,W)$ and $N(V,W)$.

Problem 13. Given a Banach space E and a measure μ, find a representation for the nonstandard hull of $L_p(\mu,E)$; can it be represented in some smooth way in terms only of $\hat{L}_p(\mu)$ and $\hat{E}$?

This is open even when E is ℓ_q and $L_p(\mu)$ is ℓ_p, although S. Heinrich [HEI 6] has identified some aspects of the structure of $\hat{F}$ where F is $\ell_p(\ell_q)$.

Problem 14. Describe the nonstandard hulls of Lorentz sequence spaces $d(w,p)$. (Here $w = (w_1,w_2,\ldots)$ is a sequence $1 = w_1 \geq w_2 \geq \ldots > 0$ with $\lim w_n = 0$ and $1 \leq p < \infty$. The norm of an element $x = (x_1,x_2,\ldots)$ in $d(w,p)$ is given by

$$\|x\| = \sup_{\pi}\left(\sum_{n=1}^{\infty} |x_{\pi(n)}|^p w_n\right)^{1/p}$$

where π ranges over all permutations of the positive integers.) These are perhaps the simplest sequence spaces for which the nonstandard hulls have not been studied in detail.

Another class of sequence spaces for which information about the nonstandard hulls is very incomplete is the class of Orlicz spaces. (See [HEI 4, Section 5].)

Problem 15. (J. Stern) Find a $\mathcal{L}_1$-space E such that no nonstandard hull of E is isomorphic to an $L_1(\mu)$-space.

Generally, consider the equivalence relation on Banach spaces defined by taking E,F to be equivalent if they have isomorphic nonstandard hulls. This is coarser than $\equiv_A$, Proposition 5.1. The results in Section 6 show that there is an equivalence class consisting of the $\mathcal{L}_p$-spaces (for each $1 < p < \infty$) and at least two equivalence classes of $\mathcal{L}_\infty$-spaces.

How many equivalence classes of $\mathcal{L}_1$-spaces and of $\mathcal{L}_\infty$-spaces are there?

Problem 16. Let C_p denote the complex Banach space of p-trace-class compact operators on the separable (infinite dimensional) Hilbert space H ($1 \leq p \leq \infty$). Find representations for the nonstandard hulls $\hat{C}_p$. Perhaps this will involve "noncommutative L_p-space theory," based on certain families of C* - algebras with traces. (See the recent expository book [SI] and its references.)

Problem 17. The theory of scalar valued Loeb measures [LO] has proved to be very important for applications of nonstandard analysis in probability theory. (For example see the survey paper by E. Perkins in this volume.) Develop a theory of vector-valued Loeb measures. In particular, if μ is a finitely additive internal measure with values in an internal Banach space V, when does μ give rise to a vector measure with values in $\hat{V}$ and which vector measures arise in this way?

Problem 18. Develop a criterion to tell when two Frechet spaces (or more general locally convex spaces) have linearly isomorphic nonstandard hulls. This may involve developing an appropriate logic for these spaces. The same problem also seems interesting for some classes of non-locally-convex topological vector spaces. See [HMR 1] for a discussion of the nonstandard hulls of such spaces in general.

Problem 19. (S. Heinrich and P. Mankiewicz) If X is uniformly homeomorphic to an $\mathcal{L}_1$-space, must X itself be an $\mathcal{L}_1$-space? (See Section 11.)

Problem 20. (S. Heinrich and P. Mankiewicz.) If E,F are standard Banach spaces which are uniformly homeomorphic, must their nonstandard hulls $\hat{E},\hat{F}$ be linearly isomorphic? What if we assume that they are super-reflexive? This question concerns the extent to which the local (isomorphic) geometry of a Banach space is determined by the uniform structure that it carries. (See Section 11.)

It should be mentioned that I. Aharoni and J. Lindenstrauss [AHL] have given examples of two (nonseparable) Banach spaces E,F such that E,F are uniformly homeomorphic but not linearly isomorphic. (On the other hand, for this pair E,F the nonstandard hulls $\hat{E},\hat{F}$ are linearly isomorphic [HMK].)

Problem 21. What does it mean for a pair E,F of standard Banach spaces that their nonstandard hulls $\hat{E},\hat{F}$ satisfy one of the hypotheses of the Pelczynski Decomposition Scheme? For example, when is $\hat{E}$ isomorphic to $\hat{E} \oplus \hat{E}$? (That is, what does this say about E?) See [LT 2] for a discussion of the Pelczynski Decomposition Method.

References.

[AHL] I. Aharoni and J. Lindenstrauss, Uniform equivalence between Banach spaces, Bull. Amer. Math. Soc. 84 (1978), 281-283.

[AML] D. Amir and J. Lindenstrauss, The structure of weakly compact sets in Banach spaces, Ann. of Math. 88 (1968), 35-46.

[BA] K. J. Barwise, Back and forth through infinitary logic, in Studies in Model Theory, M. Morley (ed.), Math. Assn. of America (Providence, 1974), 5-34.

[BEL 1] S. Bellenot, Prevarieties and intertwined completeness of locally convex spaces, Math. Ann. 217 (1975), 59-67.

[BEL 2] ______________, On nonstandard hulls of convex spaces, Canad. J. Math. 28 (1976), 141-147.

[BEN] Y. Benyamini, Separable G spaces are isomorphic to C(K) spaces, Israel J. Math. 14 (1973), 287-293.

[BNL] Y. Benyamini and J. Lindenstrauss, A predual of ℓ_1 which is not isomorphic to a C(K) space, Israel J. Math. 13 (1972), 246-254.

[BER] S. Bernau, A unified approach to the principle of local reflexivity, in Notes in Banach Spaces, H. E. Lacy (ed.), Univ. Texas Press (Austin, 1980), 427-439.

[BRR] A. Bernstein and A. Robinson, Solution of an invariant subspace problem of K. T. Smith and P. R. Halmos, Pacific J. Math. 16 (1966), 421-431.

[BDCK] J. Bretagnolle, D. Dacunha-Castelle and J.-L. Krivine, Lois stables et espaces L^p, Ann. Inst. Henri Poincaré, Sect. B 2 (1966), 231-259.

[BRS 1] Brunel and Sucheston, On B-convex Banach spaces, Math. System Theory 7 (1974), 294-299.

[BRS 2] ______________, on J-Convexity and ergodic super-properties of Banach spaces, Trans. Amer. Math. Soc. 204 (1975), 79-90.

[CON] J. Conroy, The finite dimensional Riesz subspace structure of Banach lattices and applications to the theory of nonstandard hulls, Ph.D. Thesis, Duke University, 1976.

[CNM] J. Conroy and L. C. Moore, Jr., Local reflexivity in Banach lattices, (unpublished manuscript).

[CK] H. Corson and V. Klee, Topological classification of convex sets, Proc. Symp. Pure. Math. Vol. 7 (Convexity), Amer. Math. Soc. (Providence, 1963), 37-51.

[CZM] D. Cozart and L. C. Moore, Jr., The nonstandard hull of a normed Riesz space, Duke Math. J. 41 (1974), 263-275.

[DCK 1] D. Dacunha-Castelle and J.-L. Krivine, Applications des ultraproduits a l'etude des espaces et des algèbres de Banach, Studia Math. 41 (1972), 315-334.

[DCK 2] _______________, Ultraproduits d'espaces de Banach, Sém. Goulaouic-Schwartz, 1971-1972, Exposés IX, X.

[DCK 3] _______________, Sous-espaces de L^1, Israel J. Math. 26 (1977), 320-351.

[DA] M. Davis, Applied Nonstandard Analysis, J. Wiley Pub. (New York, 1977).

[DO] H. P. Dowson, Spectral Theory of Linear Operators, Academic Press (New York, 1977).

[DS] N. Dunford and J. Schwartz, Linear Operators Part I, Interscience Publishers (New York, 1958).

[EH] A. Ehrenfeucht, An application of games to the completeness problem for formalized theories, Fund. Math. 49 (1961), 129-141.

[ELP] P. Enflo, J. Lindenstrauss and G. Pisier, On the "three space problem," Math. Scand. 36 (1975), 199-210.

[ER] P. Enflo and H. P. Rosenthal, Some results concerning $L_p(\mu)$ spaces, J. Functional Analysis 14 (1973), 325-348.

[FR] R. Fraissé, Sur quelques classifications des systèmes des relations, Publ. Sci. Univ. Alge. Ser. A 1 (1954), 35-82.

[GU] V. I. Gurarii, Spaces of universal disposition, isotropic spaces and the Mazur problem on rotations of Banach spaces, Sibirski Math. Z. 7 (1966), 1002-1013.

[HEI 1] S. Heinrich, Finite representability and super-ideals of operators, Dissertationes Math. vol. 172 (1980).

[HEI 2] _______________, Finite representability of operators, in Proc. Int. Conf. on Operator Algebras, Ideals and their Applications in Theoretical Physics, Leipzig, 1977.

[HEI 3] _______________, Closed operator ideals and interpolation, J. Funct. Analysis 35 (1980), 397-411.

[HEI 4] _______________, Ultraproducts in Banach space theory, J. Reine Angew. Math. 313 (1980), 72-104.

[HEI 5] _______________, Ultraproducts of L_1-predual spaces, Fund. Math. 113 (1981), 221-234.

[HEI 6] _______________, The isomorphic problem of envelopes, to appear.

[HH] S. Heinrich and C. W. Henson, Banach space model theory II, in preparation.

[HHM] S. Heinrich, C. W. Henson and L. C. Moore, Jr., Elementary equivalence of L_1-preduals, in preparation.

[HMK] S. Heinrich and P. Mankiewicz, Applications of ultrapowers to the uniform and Lipschitz classification of Banach spaces, preprint (Institute of Math., Polish Acad. Sci. 1980).

[HP] S. Heinrich and A. Pietsch, A characterization of (∞,p,q)-integral operators, Math. Nachr. 89 (1979), 197-202.

[HEN 1] C. W. Henson, The isomorphism property in nonstandard analysis and its use in the theory of Banach spaces, J. Symbolic Logic 39 (1974), 717-731.

[HEN 2] ________, When do two Banach spaces have isometrically isomorphic nonstandard hulls? Israel J. Math. 22 (1975), 57-67.

[HEN 3] ________, Nonstandard hulls of Banach spaces, Israel J. Math. 25 (1976), 108-144.

[HEN 4] ________, Ultraproducts of Banach spaces, in the Altgeld Book 1975-1976, The University of Illinois Functional Analysis Seminar.

[HEN 5] ________, Banach space model theory I, in preparation.

[HMR 1] C. W. Henson and L. C. Moore, Jr., The nonstandard theory of topological vector spaces, Trans. Amer. Math. Soc. 172 (1972), 405-435.

[HMR 2] ________, Nonstandard hulls of the classical Banach spaces, Duke Math. J. 41 (1974), 227-284.

[HMR 3] ________, Subspaces of the nonstandard hull of a normed space, Trans. Amer. Math. Soc. 197 (1974), 131-143.

[HMR 4] ________, The Banach spaces $\ell_p(n)$ for large p and n, to appear in Manuscripta Mathematica.

[HRB] K. Hrbacek, Axiomatic foundations for nonstandard analysis, Fund. Math. 98 (1978), 1-19.

[JA] R. C. James, Characterizations of reflexivity, Studia Math. 23 (1963/64), 205-216.

[KR 1] J.-L. Krivine, Theorie des modèles et espaces L^p, C. R. Acad. Sci. Paris Ser. A 275 (1972), 1207-1210.

[KR 2] ________, Langages à valeurs réelles et applications, Fund. Math. 81 (1974), 213-253.

[KR 3] ________, Sous espaces de dimension finie des espaces de Banach réticulés, Ann. of Math. 104 (1976), 1-29.

[KU] K. D. Kürsten, On some questions of A. Pietsch II, Teor. Funkcional. Anal. i Priloženia (Kharkov) 29 (1978), 61-73. (Russian)

[LE] H. Lemberg, Nouvelle demonstration d'un théorème de J.-L. Krivine sur la finie representation de ℓ_p dans un éspace de Banach, Israel J. Math. 39 (1981), 341-348.

[LT 1] J. Lindenstrauss and L. Tzafriri, The uniform approximation property in Orlicz spaces, Israel J. Math. 23 (1976), 142-155.

[LT 2] ________, Classical Banach Spaces I, Sequence Spaces, Ergebnisse der Math. und ihrer Grenzgebiete 92, Springer-

Verlag (Heidelberg, 1977).

[LT 3] ____________, Classical Banach Spaces II, Function Spaces, Ergebnisse der Math. und ihrer Grenzgebiete 97, Springer-Verlag (Heidelberg, 1979).

[LO] P. Loeb, Conversion from non-standard to standard measure spaces and applications in probability theory, Trans. Amer. Math. Soc. 211 (1975), 113-122.

[LUS] W. Lusky, The Gurarii spaces are unique, Arch. Math. (Basel) 27 (1976), 627-635.

[LUX 1] W. A. J. Luxemburg, A General theory of monads, in Applications of Model Theory to Algebra, Analysis and Probability, W. A. J. Luxemburg (ed.), Holt, Rinehart and Winston (New York, 1969), 18-86.

[LUX 2] ____________, On some concurrent binary relations occurring in analysis, in Contributions to Non-Standard Analysis, W. A. J. Luxemburg and A. Robinson (eds.), North-Holland (Amsterdam, 1972), 85-100.

[LUX 3] ____________, Notes on Banach function spaces XVIB, Proc. Acad. Sci. Amsterdam A68 (1965), 658-667.

[LXZ] W. A. J. Luxemburg and A. C. Zaanen, Riesz Spaces I, North-Holland (Amsterdam, 1971).

[MN] P. Meyer-Nieberg, Characterisierung einiger topologischer und ordnungstheoretischer Eigenschaften von Banachverbänden unit Hilfe disjunkter Folgen, Arch. Math. (Basel) 24 (1973), 640-647.

[MO 1] L. C. Moore, Jr., Hyperfinite extensions of bounded operators on a separable Hilbert space, Trans. Amer. Math. Soc. 218 (1976), 285-295.

[MO 2] ____________, Hyperfinite-dimensional subspaces of the non-standard hull of c_0, Proc. Amer. Math. Soc. 80 (1980), 597-603.

[MO 3] ____________, Approximately finite-dimensional Banach spaces, J. Funct. Analysis 42 (1981), 1-11.

[MO 4] ____________, Unitary equivalence of nonstandard extensions of bounded operators on Hilbert space, preprint.

[NE] E. Nelson, Internal set theory: A new approach to nonstandard analysis, Bull. Amer. Math. Soc. 83 (1977), 1165-1198.

[PR] A. Pelczynski and H. P. Rosenthal, Localization techniques in L_p-spaces, Studia Math. 52 (1975), 263-289.

[PT] A. Pietsch, Ultraprodukte von Operatoren in Banachräumen, Math. Nachr. 61 (1974), 123-132.

[RA] S. A. Rakov, C-convexity and the "three space problem," Dokl. Acad. Nauk. SSSR 228 (1976), 303-305.

[RI 1] M. Ribe, On uniformly homeomorphic normed spaces, Ark. Math. 14 (1976), 237-244.

[RI 2] __________, On uniformly homeomorphic normed spaces, II, Ark. Math. 16 (1978), 1-9.

[ROB 1] A. Robinson, On generalized limits and linear functionals, Pacific J. Math. 14 (1964), 269-283.

[ROB 2] __________, Non-Standard Analysis, North-Holland (Amsterdam, 1966).

[RZ] A. Robinson and E. Zakon, A set-theoretical characterization of enlargements, in Applications of Model Theory to Algebra, Analysis and Probability, W. A. J. Luxemburg (ed.), Holt, Rinehart and Winston (New York, 1969), 109-122.

[ROS 1] H. P. Rosenthal, A characterization of Banach spaces containing ℓ_1, Proc. Nat. Acad. Sci. (USA), 71 (1974), 2411-2413.

[ROS 2] __________, On a theorem of J. L. Krivine, J. Funct. Analysis 28 (1978), 197-225.

[SI] B. Simon, Trace Ideals and Their Applications, London Math. Soc. Lecture Notes 35, Cambridge Univ. Press (Cambridge, 1979).

[STE 1] J. Stern, Sur certaines classes d'espaces de Banach caracterisees par des formules, C. R. Acad. Sci. Paris Ser. A 278 (1974), 525-528.

[STE 2] __________, Some applications of model theory in Banach space theory, Ann. Math. Logic 9 (1976), 49-121.

[STE 3] __________, The problem of envelopes for Banach spaces, Israel J. Math. 24 (1976), 1-15.

[STE 4] __________, Ultrapowers and local properties of Banach spaces, Trans. Amer. Math. Soc. 240 (1978), 231-252.

[STL] K. D. Stroyan and W. A. J. Luxemburg, Introduction to the Theory of Infinitesimals, Academic Press (New York, 1976).

[ZI] M. Zippin, On perfectly homogeneous bases in Banach spaces, Israel J. Math. 4 (1966), 265-272.

STOCHASTIC SOLUTIONS TO PARTIAL DIFFERENTIAL EQUATIONS

S. A. Kosciuk
University of Wisconsin
Madison, WI 53706

It is well known that solutions of certain diffusion equations can be realized as functionals of stochastic processes. For example let

$$L\phi = \frac{1}{2}\,\mathrm{trace}(aH\phi) + b\cdot\nabla\phi$$

where $a(t,x)$ is a symmetric non-negative definite dxd real matrix, $b(t,x) \in R^d$, $H\phi$ and $\nabla\phi$ are the Hessian matrix and gradient of ϕ in the space variables $x \in R^d$. Then

$$u(t,x) = E^{p_x}\Big[\int_t^T f(s,x(\omega,s))ds\Big] \quad \text{solves}$$

1) $$u_t + Lu = -f \quad \text{on } [o,T] \times R^d$$

where p_x is a probability measure and $x(\omega,s)$ is a stochastic process such that for all smooth ϕ

2) $$\phi(t,x(\omega,t)) = \int_0^t \Big(\frac{\partial}{\partial s} + L\Big)\phi(s,x(\omega,s))ds \quad \text{is a martingale}$$

and $x(\omega,o) = x$ a.s. p_x.

The sense in which 1) holds depends on the properties of a and b. For example if a and b are smooth enough then u will be smooth and 1) will hold in the ordinary sense. However, if a is not continuous and allowed to degenerate then there may be no family of measures $\{p_x : x \in R^d\}$ such that even 2) is satisfied.

In this paper general conditions on a and b are given which allow 2) to be solved for a family of measures $\{p_x : x \in R^d\}$ when a is allowed to be discontinuous and degenerate.

The interest in such diffusions lies in constructing models for systems such as DNA-enzyme reactions in which enzymes diffuse randomly through space and then in a one-dimensional fashion along the DNA molecule. Deterministic models lead to non-linear reaction-diffusion equations in the absence of excess reactants, whereas stochastic models can be realized using linear but random generators.

A family of measures $p_{s,x}$ and a stochastic process $x(t)$ on an adapted probability space (Ω, F_t) for which 2) is an F_t-martingale under $p_{s,x}$ and $x(\omega,s) = x$ a.s. $p_{s,x}$ is said to solve the martingale problem (MGP) for a,b starting at (s,x).

We will use the following estimate due to N.V. Krylov. Let S^d be the symmetric non-negative definite dxd real matrices.

Th. Let (Ω, F_t, p) be an adapted probability space and $a: \Omega x[0,1] \to S^d$ and $b: \Omega x[0,1] \to R^d$ be bounded and progressively measurable.

Let $L(\omega,t)\phi = \frac{1}{2}\,\mathrm{trace}(a(\omega,t)H\phi) + b(\omega,t)\cdot\nabla\phi$

Let $x: \Omega x[0,1] \to R^d$ be progressively measurable such that for all smooth ϕ

$$\phi(t,x(\omega,t)) - \int_0^t \left(\frac{\partial}{\partial s} + L(\omega,s)\right)\phi(s,x(\omega,s))ds \text{ is an } F_t\text{-}$$

martingale under p.

Then for all measurable f

$$E^p\left[\int_0^1 (\det(a(\omega,t)))^{1/d+1} f(t,x(\omega,t))\right] \le C\,\|f\|_{d+1}$$

where C depends only on d and the bounds for a and b, and $\|\ \|_{d+1}$ is the L^{d+1}-norm on $[0,1] \times R^d$.

We will use the familiar notation of non-standard analysis. ${}^*A$ is the *-extension of the standard set A in the non-standard universe, $x \approx y$ means "x is infinitely close to y", and $°x$ denotes the unique standard object (if one exists) such that $x \approx °x$. The topology defining $\approx$ and ° will be clear from the context.

Our probability spaces, Ω, will be internal sets of paths $\omega: T \to {}^*R^d$ for $T = \{0, \Delta t, 2\Delta t, \ldots, 1\}$ $\Delta t \approx 0$, $\Delta t > 0$. For $t \in [0,1]$ we define A_t to be the external σ-algebra of subsets of Ω generated by sets closed under the equivalence relation $\sim_t$ where $\omega \sim_t \omega'$ if for each $s \in T$ with $s < t$ or $s \approx t$ $\omega(s) = \omega'(s)$. We will solve the MGP on the adapted spaces (Ω, A_t), that is we will construct A_t-martingales.

For $\omega \in \Omega$, $t \in T$ write $\omega\restriction_t$ for $\omega \restriction \{0, \Delta t, \ldots, t\}$ and set $(\omega\restriction t) = \{\alpha \in \Omega: \alpha\restriction_t = \omega\restriction_t\}$. Let p be an internal probability weighting on Ω ie. $p(\omega) \ge 0$ and $\sum_{\omega \in \Omega} p(\omega) = 1$. For $s,t \in T$ set

$$E^p[X(\omega,s)\,|\,\omega\restriction_t] = (p(\omega\restriction_t))^{-1} \sum_{\alpha \in (\omega\restriction t)} X(\alpha,s) \quad \text{where}$$

$X: \Omega x T \to {}^*R^d$ and $p(\omega) > 0$ for at most a hyperfinite number of ω.

References:

For further reading on the martingale problem and differential equations a good source is:

D.W. Stroock, S.R.S. Varadan, "Multidimensional Diffusion Processes" (Springer-Verlag, New York, 1979).

For Krylov's theorem consult,

N.V. Krylov, "Controlled Diffusion Processes" (Springer-Verlag, New York, 1980).

N.V. Krylov, Sequences of convex functions and estimates of the maximum of the solution of a parabolic equation, Siberian Math J. 17(2) (1976); 226-236 (English translation.

For non-standard probability consult

H.J. Keisler, An infinitesimal approach to stochastic analysis, 1980 (yet to be published).

Existence of Solutions to the Martingale Problem in R^d for Degenerate Discontinuous Coefficients

Theorem:

Let S^d be the symmetric non-negative definite real matrices. Let $a:[0,1]\times R^d \to S^d$ and $b:[0,1]\times R^d \to R^d$ be bounded and measurable. Let $D = \{(t,x): \det a(t,x) = 0\}$

Assume that a and b are continuous on D. Assume also that if $(t,x) \in D$ is not a point of continuity of both a and b then there is an open U with $(t,x) \in U$ such that the det a is bounded away from zero on $U \setminus D$.

Then for each $(s,x) \in [0,1] \times R^d$ there exists a solution to the Martingale Problem (MGP) for a,b starting at (s,x).

<u>Proof</u> for clarity we assume that $b(\cdot,\cdot) = 0$, $a(\cdot,x) = a(x)$, a has compact support, and that a is bounded so that trace $a \leq 1$. Let $S_1^d = \{a \in S^d : \mathrm{tr}\, a \leq 1\}$ also for $A \subset R^d$ let $A^c = R^d \setminus A$.

<u>Step 1</u> Let $\underline{a}: {}^*R^d \to {}^*S_1^d$ be internal. We construct a solution of a hyper-martingale problem for $\underline{a}$. Let $K \in {}^*N \setminus N$ set $\Delta t = \frac{1}{K}$ and let $T = \{0, \Delta t, 2\Delta t, \ldots, 1\}$. Let $\Omega = \{\omega : T \to {}^*R^d \mid \omega$ is internal$\}$.

For each $z \in {}^*R^d$ we define a weighting $p_{\underline{a}}^z$ on Ω. We first define the "transition probability" function $p(x,y)$ which will represent the probability under $p_{\underline{a}}^z$ that a path $\omega \in \Omega$ will be at y at time $t+\Delta t$ given that it is at x at time t. Let $\lambda^i(x)$, $v^i(x)$ $1 \leq i \leq d$ be the eigenvalues and orthonormal eigenvectors of $\underline{a}(x)$. If some $\lambda^i(x) = 0$ we still choose an orthonormal basis $V^i(x)$ $1 \leq i \leq d$. Let $\Delta x = \sqrt{\Delta t}$.

Define

$$p(x,y) = \begin{cases} \frac{1}{2}\lambda^i(x) & \text{if } y = x \pm \Delta x v^i(x) \quad 1 \leq i \leq d \\ 1 - \sum_i \lambda^i(x) & \text{if } y = x \\ 0 & \text{otherwise} \end{cases}$$

Define

$$p_{\underline{a}}^z(\omega) = \begin{cases} \prod_{t \in T} p(\omega(t), \omega(t+\Delta t)) & \text{if } \omega(0) = z \\ 0 & \text{otherwise} \end{cases}$$

Then $\sum_{\omega\varepsilon\Omega} p_{\underline{a}}^z(\omega) = 1$ so $p_{\underline{a}}^z$ is an internal probability.

We now show that $p = p_{\underline{a}}^z$ solves an appropriate hypermartingale problem for $\underline{a}$.

For f internal let

$$L^{\Delta}f(x) = \frac{1}{2} \sum_{1\leq i\leq d} \lambda^i(x)(\Delta x)^{-2}(f(x+\Delta x v^i(x)) + f(x-\Delta x v^i(x)) - 2f(x)).$$

A straightforward calculation shows that

$E^p[f(\omega(t+\Delta t)) - f(\omega(t))|\omega\restriction_t] = L^{\Delta}f(\omega(t))\Delta t$, but this means that

$f(\omega(t)) - \sum_{s=0}^{t-\Delta t} L^{\Delta}f(\omega(s))\Delta t$ is an $(\omega\restriction t)$-martingale under p.

Note also that if $a:R^d \to S_1^d$ and $\underline{a}(x) \approx a({}^\circ x)$ and f is smooth then $L^{\Delta^*}f(x) \approx \frac{1}{2}$ trace$(aHf({}^\circ x))$ when Hf is the Hessian matrix of f. Let $Lf = \frac{1}{2}$ trace(aHf). Also for $\omega\varepsilon\Omega$ let $^\circ\omega$ denote that function from $[0,1]$ into R^d such that $(\forall t\varepsilon T)\,{}^\circ(\omega(t)) = {}^\circ\omega({}^\circ t)$. Since $\underline{a}$ is bounded $^\circ\omega$ is defined and continuous for $^\circ p$ a.a.ω.

Step 2. We now define an internal $\underline{a}: {}^*R^d \to {}^*S_1^d$ s.t. $^\circ(p_{\underline{a}}^z)$ solves the MGP for a starting at $^\circ z$. We choose a sequence $a_n \to a$ a.s. in x and $\underline{a}$ will be an a_H for some infinite H. We also require

0) a_n is continuous and $a_n(x) \varepsilon S_1^d$

1) $\exists U_n \subset R^d$ open and bdd. with $a_n = a$ off U_n, $U_{n+1} \subset U_n$, & $mU_n < \frac{1}{n}$ (Here m is Lebesque measure on R^d)

2) $(\forall n)(\forall \ell\leq n)(\exists W_\ell^n, V_\ell^n)[W_\ell^n, V_\ell^n$ are open &

$W_\ell^n \subset \bar{W}_\ell^n \subset V_\ell^n \subset \bar{V}_\ell^n \subset U_\ell$ & $W_\ell^{n+1} \subset W_\ell$ & $V_\ell^{n+1} \subset V_\ell^n]$

3) $a_n = \frac{1}{d} I$ on W_n^n

4) $(\exists f\downarrow o)(\forall n)(\forall \ell\leq n)(\forall x,y\varepsilon \bar{U}_\ell \setminus V_\ell^n)[(|y-x|\leq f(n) \to |a_n(y)-a_n(x)| \leq \frac{1}{n}$ & dist$(x,U_\ell^c) \leq f(n)]$

5) $(\exists g\downarrow o)(\forall n)(\forall k\leq n)(\forall x\varepsilon U_n \setminus W_n^n)[\det a_n(x) < \frac{1}{k} \to (\exists y\varepsilon U_n^c)(|y-x| < \frac{1}{k}$ & $|a_n(y) - a_n(x)| < g(n)]$

Assuming $\{a_n : n\varepsilon N\}$ is such a sequence consider the *-sequence $\{{}^*a_n : n\varepsilon {}^*N\}$ for $H\varepsilon {}^*N\setminus N$ let $B_H = \{x\varepsilon {}^*R^d : a_H(x) \neq a({}^\circ x)\}$. Then by 0) and 1) $B_H \subset \bigcap_{\ell<\infty} {}^*U_\ell$ and by 2) and 4) $B_H \subset \bigcap_{\ell<\infty} V_\ell^H$.

We now define the a_n leaving the reader to prove properties 0)-5).

Since we assume that a has compact support by Lusin's theorem we can choose U_n s.t. 1) holds i.e., a is continuous off U_n. What remains is to define a_n on U_n. Let h_n represent the modulus of continuity of a off U_n i.e., $(\forall n,m)(\forall x,y\varepsilon U_n^c)[|x-y|\le h_n(m)\to|a_n(x)-a_n(y)|<\frac{1}{3m}]$. Let $\hat{f}(n) = h_n(m_n)$ where $h_n(m_n)\downarrow 0$ and $m_n\uparrow\infty$ as $n\uparrow\infty$ also $m_n \ge n$. Set $W_\ell^n = \{x\varepsilon R^d : d(x,U_\ell^c)>\hat{f}(n)\}$, let $f(n) = (1/5)\hat{f}(n)$ and set $V_\ell^n = \{x\varepsilon R^d : D(x,U_\ell^c)>f(n)\}$ let $\tilde{a}_n(x) = (1/d)I$ on $\bar{W}_n^n$ and $a(x)$ off U_n let $r_n(x) = d(x,U_n^c)\wedge d(x,W_n^n)$ let $B_x^n = \{y:|y-x|\le r_n(x)\}$. Let $\hat{a}_n$ be measurable s.t. $\hat{a}_n(x)\varepsilon\{\hat{a}(y):y\varepsilon B_x^n\}\subset S_1^d$. Define $a_n(x) = (1/mB_x^n)\int_{B_x^n}\hat{a}_n(y)dy$ if $r_n(x)\ne 0$ and $a_n(x) = \hat{a}(x)$ o.w.

Then $a_n(x)$ is defined for all $x\varepsilon R^d$ and 0) through 5) hold.

<u>Step 3</u> Let λ be counting measure on T. Set $p_n = p^z_{a_n}$ $n\varepsilon {}^*N$. To show that ${}^\circ p_H$ solves the MGP for a $(H\varepsilon {}^*N/N)$ it suffices to show that except for a ${}^\circ(p_H\times\lambda)$ null set for smooth f

$$^*f(\omega(t)) - \sum_{s=0}^{t-\Delta t} L^{\Delta *}f(\omega(s))\Delta t \approx f({}^\circ\omega({}^\circ t)) - \int_0^{{}^\circ t} Lf({}^\circ\omega(s))ds.$$

Since the l.h.s. is an $(\omega\uparrow t)$-martingale for p_H the r.h.s. will be an A_t-martingale for ${}^\circ p_H$ by Hoover's and Keisler's lifting theorem. Hence we must show

$$(*)\quad {}^\circ(p_H\times\lambda)\{(\omega,t):a_H(\omega(t)) \ne a({}^\circ\omega({}^\circ t))\} = 0$$

for sufficiently small infinite H. Since $\forall n\varepsilon {}^*N$ the paths $\omega(\cdot)$ are p_n a.s. S-continuous it suffices to show

$$(*)'\quad {}^\circ(p_H\times\lambda)\{(\omega,t):a_H(\omega(t)) \ne a({}^\circ(\omega(t)))\} = {}^\circ(p_n\times\lambda)\{(\omega,t):\omega(t)\varepsilon B_H\} = 0$$

Note that for $n\varepsilon N$ since $B_n = \phi$ (i.e., ${}^*a_n(x)\approx a_n({}^\circ x)$ $x\varepsilon {}^*R^d$) ${}^\circ p_n$ solves the MGP for a_n.

For $H\varepsilon {}^*N$ let $D_H = \{x\varepsilon {}^*R^d:\det a_H(x)\approx 0\}$. We first prove the following

<u>Claim</u>: $(\forall H\varepsilon {}^*N\setminus N)[B_H\cap D_H = \phi]$

<u>pf</u> fix $H\varepsilon {}^*N\setminus N$ we first show ${}^\circ D_H\subset D = \{x\varepsilon R^d:\det a(x) = 0\}$. Under the continuity assumptions on a $D = \{x:(\exists y_k\to x)[\det a(y_k)\to 0]\}$. Fix $x\varepsilon D_H$, we define $(y_k)\subset R^d$ $y_k\to{}^\circ x$ & $\det a(y_k)\to 0$. Now for any $x\ \varepsilon {}^*R^d$ $\exists(n_k)\subset N$ and $(x_k)\subset R^d$ s.t. $x_k\to{}^\circ x$ & $a_{n_k}(x_k)\to{}^\circ a_H(x)$. By

passing to a subsequence of the (n_k) and (x_k) assume that $n_k \geq k$ and $\det a_{n_k}(x_k) < \frac{1}{k} < (\frac{1}{d})^d$. If $x_k \notin U_{n_k}$ set $y_k = x_k$ otherwise $x_k \in U_{n_k} \setminus W_{n_k}^{n_k}$ (i.e., $x \in W_n^n \to \det a_n(x) = (\frac{1}{d})^d$) so by property 5) from step 2 $\exists y \in U_{n_k}^c$ s.t. $|y-x_k| < \frac{1}{k}$ and $|a_{n_k}(y) - a_{n_k}(x_k)| < g(n_k)$ set $y_k = y$ in this case. Then $y_k \to {}^\circ x$ and $\lim a_{n_k}(y_k) = \lim a_{n_k}(x_k) = {}^\circ a_H(x)$ but $a_{n_k}(y_k) = a(y_k)$ since $y_k \notin U_{n_k}$ so $\det a(y_k) \to \det {}^\circ a_H(x) = 0$ ie ${}^\circ x \in D$. Now if ${}^\circ x$ is a point of continuity of a then $a(y_k) \to a({}^\circ x)$ and from above $a(y_k) \to {}^\circ a_H(x)$ so ${}^\circ a_H(x) = a({}^\circ x)$ i.e., $x \notin B_H$. Otherwise $\exists U \ni x$ open and $\varepsilon > 0$ s.t. $\det a > \varepsilon$ on $U \setminus D$. But $\det a(y_k) < \varepsilon \rightarrow y_k \in D$ so $\lim a(y_k) = a({}^\circ x)$ since a is continuous on D, as before $a({}^\circ x) = {}^\circ a_H(x)$ so $x \notin B_H$.

We now prove (*)'. For $n,k \in {}^*N$ and $A \subset {}^*R^d$ set $d_k^n(A) = \left\{x \in A : \det a_n(x) > \frac{1}{k}\right\}$ since $B_H \cap D_H = \emptyset$ and $B_H \subset \bigcap_{\ell<\infty} V_\ell^H$ we see $B_H \subset \bigcup_{k<\infty} \bigcap_{\ell<\infty} d_k^H(V_\ell^H)$. Since $d_k^H(V^H)$ is internal

$$ {}^\circ(p_H^z \times \lambda)\left\{(\omega,t) : \omega(t) \in B_H\right\} \leq \lim_k [\lim_\ell {}^\circ((p_H^z \times \lambda)\left\{(\omega,t) : \omega(t) \in d_k^H(V_\ell^H)\right\})]. $$

Hence it suffices to show there is a constant, C, independent of n, k, ℓ s.t.

$$ (*)'' \quad (\forall \ell,k,n \in N)[(p_n^z \times \lambda)\left\{(\omega,t) : \omega(t) \in d_k^n({}^*V_\ell^n)\right\} \leq \frac{k \cdot C}{\ell}]. $$

$\forall \ell \leq n < \infty$ let ρ_ℓ^n be smooth s.t. $\rho_\ell^n = 0$ off U_ℓ, $\rho_\ell^n = 1$ on V_ℓ^n and $0 \leq \rho_\ell^n \leq 1$. Then since a_n is continuous and ${}^\circ(p_n^z)$ solves the MGP for a_n by Krylov's theorem we have

$$ (p_n^z \times \lambda)\left\{(\omega,t) : \omega(t) \in d_k^n({}^*V_\ell^n)\right\} \leq kE^{p_n^z}[\sum_0^1 \det {}^*a_n((t))\, {}^*\rho_\ell^n(\omega(t)\Delta t] $$

$$ \approx kE^{{}^\circ p_n^z}[\int_0^1 \det a_n({}^\circ\omega(s))\rho_\ell^n({}^\circ\omega(s))ds] $$

$$ \leq k \cdot C \, \|\rho_\ell^n\|_{d+1} \leq \frac{k \cdot C}{\ell} \text{ since } mU_\ell < \frac{1}{\ell}. $$

This completes the proof.

Ω-Calculus as a Generalization of Field Extension
An Alternative Approach to Nonstandard Analysis

by Detlef Laugwitz

Technische Hochschule Darmstadt
Fachbereich Mathematik
Schloßgartenstr. 7
D - 6100 Darmstadt

Introduction

The main features of the approach to infinitesimals and infinitely large numbers which is proposed here are the following:

Let T be some theory which contains an elementary theory of natural and rational numbers. An extended theory $T\langle\Omega\rangle$ is obtained by adding a new extra logical constant Ω which, in a precise way, behaves like very large natural numbers. In particular, Ω is larger than every natural number n_o of T. If T is a theory of real analysis containing the ordered field axioms for $\mathbb{R}$, then these axioms are true for the numbers of the "non standard" theory $T\langle\Omega\rangle$.

It may happen that, by our Basic Definition, a sentence $A \vee B$ is true in $T\langle\Omega\rangle$, but that we are unable to decide whether A or B or both are true in $T\langle\Omega\rangle$. Though this is an unusual situation it does not diminish the effectiveness of our method. Advantages are that the approach is rather direct, and that neither tools from mathematical logic nor ultrapowers are needed.

On the other hand $T\langle\Omega\rangle$ is - in a precise sense - equivalent to a nonstandard theory. We shall show that in section 6.

It is a pleasure for me to acknowledge the great influence that discussions with C. Schmieden have had on the results of the present paper.

1. The Basic Definition

Let a theory T be given which is a collection of theorems or true statements formulated in a language L. The expressions of L are formed from an alphabet A, which in any case will contain the equality sign $=$, the logical connectives $\neg, \wedge, \vee, \Rightarrow, \Leftrightarrow$, the logical quantifiers $\bigvee, \bigwedge$, symbols for extralogial constants and variables, and notational symbols like brackets.

As an example for T take some elementary arithmetic of the rational numbers. If T is some classical theory like Gauss' Disquisitiones Arithmeticae or any constructive theory the above alphabet will presumably suffice. The extra-logical symbols will describe numbers, functions, operations, and more general relations.

For other theories, the alphabet will have to be extended. If T includes a naive or some axiomatic set theory, we shall add symbols from set theory, $\in, \subseteq, \cap, \ldots$ and symbols for set constants and variables, for sets of sets, etc. The reader may think of the axioms of Peano and ZFC as true sentences of T.

The particular kind of theory T is not important for our extension procedure, and the choice of T is left to the taste of the reader. We assume that A contains number constants and number variables for the integers and the rationals, and that T contains the basic theorems on integers and rationals, in particular the ordered field properties of the rationals and the axiom or theorem of complete induction for natural numbers.

From A and L we obtain a new alphabet $A\langle\Omega\rangle$ and a new language $L\langle\Omega\rangle$ by adding a new number constant Ω to A. This leads to a new theory $T\langle\Omega\rangle$ as a collection of statements for which we shall define the notion of "true".

Basic Definition: Let $S(n)$ *for each* $n = 1,2,3,\ldots$ *be a statement formulated in* L. *If* $S(n)$ *is true for almost all* n, *then by definition* $S(\Omega)$ *will be true.*

In other words: If a natural number n_o exists, such that for the particular predicate $S(.)$ we know that $S(n)$ belongs to the collection T for $n \geq n_o$, then $S(\Omega)$ belongs to the new theory $T\langle\Omega\rangle$. As an abbreviation for the clumsy expression "for almost all natural numbers n the following is true (belongs to T)" we shall simply use the expression (fan). Our Basic Definition now reads:

(BD) If (fan) $S(n)$, then $S(\Omega)$ is true (belongs to $T\langle\Omega\rangle$).

Note that neither (fan) nor "If ..., then ..." belong to our formal languages, but are expressions from the metalanguage which we use in speaking about the theories.

We observe that (BD) is a generalization of the well known extension procedure in the algebra of ordered fields. Let K be an ordered archimedean field, where the extra logical constants are usually restricted to elements of K, the rational operations, the order relation, and where the variables are ranging over elements of K, and the predicates S are restricted to those occuring in the axioms of an ordered field. Then (BD) leads to the theory of the ordered field $K(\Omega)$ of the rational expressions in Ω with coefficients from K. Note that $\Omega > n_o$ for all natural numbers n_o since $(\text{fan})n > n_o$. Of course, in this simple example the definition (BD) is not explicitly needed since we have a model for which the properties of an ordered field are easily verified.

2. <u>Ω-rational numbers</u>

We assume that T is some specified theory including an elementary arithmetic of rational numbers. Let $a,b,c,\ldots$ denote sequences of rational numbers. By (BD) we obtain

$$\text{If} \quad (\text{fan})\ a(n) = b(n), \quad \text{then} \quad a(\Omega) = b(\Omega) \quad \text{is true.}$$

Immediate consequences of this definition of equality are its symmetry, reflexivity and, in particular, its transitivity:

Since $(\text{fan})\ a(n) = b(n) \wedge b(n) = c(n) \Rightarrow a(n) = c(n)$,
the following is true:
$a(\Omega) = b(\Omega) \wedge b(\Omega) = c(\Omega) \Rightarrow a(\Omega) = c(\Omega)$.

The equivalence classes of $=$ will be called Ω-rational numbers. We shall use Greek letters $\alpha,\beta,\gamma,\ldots$ to denote the classes of $a(\Omega)$, $b(\Omega)$, $c(\Omega)$, ...; it will be useful to write $a(\Omega) = \alpha$. From historical reasons, the letter Ω will be reserved to denote the particular Ω-rational number which is generated by the identity map $i(n) = n$:

$$\text{If} \quad (\text{fan})\ i(n) = n, \quad \text{then} \quad i(\Omega) = \Omega.$$

<u>Proposition:</u> The axioms of a commutative ordered field are true for the Ω-rational numbers.

<u>Proof:</u> The axioms are directly verified, in particular the possibi-

lity of division and the order properties:

Since (fan) $a(n) = 0 \vee \bigvee_{c(n)} a(n)c(n) = b(n)$

the following is true:

$\alpha = 0 \vee \bigvee_{\gamma} \alpha\gamma = \beta;$

Since (fan) $a(n) > 0 \vee a(n) = 0 \vee a(n) < 0$

the following is true:

$\alpha > 0 \vee \alpha = 0 \vee \alpha < 0.$

Remarks: The archimedean property is true in the following sense, where the $m(n)$ are natural numbers and $\mu = m(\Omega)$ is accordingly called an Ω-natural number:

Since (fan) $0 < a(n) < b(n) \Rightarrow \bigvee_{m(n)} b(n) \leq m(n)a(n)$

the following is true:

(A) $0 < \alpha < \beta \Rightarrow \bigvee_{\mu} \beta \leq \mu\alpha$

An Ω-rational number $\alpha = a(\Omega)$ will be called standard if $a(n)$ can be chosen such that (fan) $a(n) = a_0$. In this case we shall identify α with the rational number a_0, $\alpha = a_0$. The restricted archimedean property says that the number μ in (A) can be chosen as a standard natural number, $\mu = m_0$. This property does obviously not hold for the Ω-rational numbers. An example is $\alpha = 1$, $\beta = \Omega$. We shall call an Ω-rational number ρ infinitely large, if $\rho > m_0$ for each standard m_0. An example for an infinitely large number is Ω. The Ω-rational numbers are non-archimedean in the usual sense of the word.

A number α will be called infinitesimal if $|\alpha| < \frac{1}{m_0}$ for each standard natural number m_0. An example is $\omega = \frac{1}{\Omega}$. We shall write $\alpha \approx 0$ if α is infinitesimal, and $\alpha \approx \beta$ if $\alpha - \beta \approx 0$. An example which displays some of the notational consequences of our Basic Definition (BD) is

$$(1 + \frac{1}{\Omega})^{\Omega} \approx \sum_{k=0}^{\Omega} \frac{1}{k!} .$$

This can be proved standardly, though a proof in a nonstandard setting would display some interesting features.

It should be stressed that the present extension is different from that of our earlier paper [2] where we obtained a partially ordered ring and not an ordered field. The procedure of the earlier paper amounts to applying the Basic Definition (BD) to such formulas $S(n)$ only which do not contain logical symbols, and it does not apply to predicates. It follows that, a fortiori, all of the results of [2] are true in the present extension. Moreover, some essential features of Schmieden's ideas in [2] keep valid in our extension, in particular the explicit representation of Ω-numbers by classes of sequences of standard numbers, and, on the other hand, the possibility of a direct calculus with expressions containing Ω.

One might suspect that there mere adjoining of a single new constant Ω could not lead to a theory comparable to that of a usual nonstandard $^*\mathbb{Q}$. Surprisingly enough, we shall be able to show in section 6 that our theory is equivalent to that of a $^*\mathbb{Q}$. Before that we shall give some more or less simple examples to show how our Basic Definition is working.

3. Some Simple Lemmas on Sets and Sequences

Let T be any theory which contains an elementary number theory of the rationals as well as some sufficient set theory. We consider sequences of objects or entities $R(n)$, $S(n)$, ... from T, which may be numbers, sets, functions, etc. The Basic Definition (BD) yields for the objects of $T\langle\Omega\rangle$:

$$\text{If (fan) } R(n) = S(n), \text{ then } R(\Omega) = S(\Omega).$$

For the sake of brevity we shall frequently use the corresponding Greek letter, $P = R(\Omega)$, $\Sigma = S(\Omega)$. The properties of an equality are proved as in section 1. If all of the $S(n)$ are of the same type, say numbers or sets or functions, then Σ will be called an Ω-number or an internal set or an internal function.

As any relation, the ϵ of set theory extends in a natural way:

$$\text{If (fan) } x(n) \in S(n), \text{ then } \xi \in \Sigma,$$

where $\Sigma = S(\Omega)$ and $\xi = x(\Omega)$. Two internal sets Σ_1, Σ_2 are equal if and only if they have the same elements. This is a consequence of (BD):

Since (fan) $[x(n) \in S_1(n) \Leftrightarrow x(n) \in S_2(n)] \Leftrightarrow S_1(n) = S_2(n)$ the following is true:

$$[\xi \in \Sigma_1 \Leftrightarrow \xi \in \Sigma_2] \Leftrightarrow \Sigma_1 = \Sigma_2 .$$

If for an object $\Sigma = S(\Omega)$ we have that (fan) $S(n) = S$ the object will be called the Ω-extension of S. Sometimes Σ is called standard. From our Basic Definition we obtain

$$(\text{fan})\ S(n) = S \quad \text{yields} \quad \Sigma = S.$$

Though this is correct the notations may be misleading in the case of sets since S in $T\langle\Omega\rangle$, considered externally, may be larger than the corresponding S of the original theory T. Actually this always happens when S is an infinite set. Sometimes we shall follow the usual convention to write $^*S$ for S as considered as an object of $T\langle\Omega\rangle$. Inside $T\langle\Omega\rangle$ there can be no confusion at all since the collection of those $x(\Omega)$ which are equal to some x, where $x \in S$ in T, is not an internal set. The star $*$ is only needed to distinguish a set of T, say $\mathbb{N}$, from its Ω-extension $^*\mathbb{N}$ in $T\langle\Omega\rangle$. Externally, speaking about $T\langle\Omega\rangle$, $\mathbb{N}$ without $*$ means the collection, or external set, of those $m_o = m(\Omega) \in {}^*\mathbb{N}$ for which (fan) $m(n) = m_o$. This is, of course, the usual situation in nonstandard analysis.

An important class of theories T are the superstructures S of a set $S \supseteq \mathbb{N}$ of individuals (Urelemente) where set theory may mean ZFC. The reader is referred to the literature [1], [3]. We want to stress that our extension procedure is by no means confined to these examples.

We shall need internal sequences of objects or entities in $T\langle\Omega\rangle$, $\beta = \alpha(\mu)$, $\mu \in {}^*\mathbb{N}$. This means that for some double sequence $a(n,m)$ and some sequence $b(n)$ of objects in T we have $b(\Omega) = a(\Omega, m(\Omega))$ where $b(\Omega) = \beta$, $m(\Omega) = \mu$, $a(\Omega, m(\Omega)) = \alpha(\mu)$. If $a(k,m) = c(m)$ does not depend on k then $a(\Omega, m(\Omega)) = c(m(\Omega))$ is a standard sequence, for which we write $\gamma(\mu)$. Here $\gamma(m)$ is standard for standard natural m.

To show how our method of proof works we give some examples.

<u>Lemma 1:</u> Each non-empty internal set $\Sigma \subseteq {}^*\mathbb{N}$ contains a smallest element.

<u>Proof:</u> Let $S(n)$ be a sequence such that $\Sigma = S(\Omega)$. Since

$$(\text{fan})\ S(n) = \emptyset \vee \bigvee_{a(n)} [a(n) \in S(n) \wedge \bigwedge_{b(n)} b(n) \in S(n) \Rightarrow a(n) \leqq b(n)]$$

the following is true in $T\langle\Omega\rangle$

$$\Sigma = \emptyset \vee \bigvee_{\alpha} [\alpha\in\Sigma \wedge \bigwedge_{\beta} \beta\in\Sigma \Rightarrow \alpha \leqq \beta . \quad \Box$$

<u>Lemma 2:</u> (Induction) If an internal set $\Sigma \subseteq {}^{*}\mathbb{N}$ contains the number 1, and if $\mu+1 \in \Sigma$ follows from $\mu\in\Sigma$, then $\Sigma = {}^{*}\mathbb{N}$.

<u>Proof:</u> Since

$$(\text{fan})\ [1 \in S(n) \wedge \bigwedge_{m(n)} m(n) \in S(n) \Rightarrow m(n) + 1 \in S(n)] \Rightarrow S(n) = \mathbb{N}$$

we conclude

$$[1\in\Sigma \wedge \bigwedge_{\mu} \mu\in\Sigma \Rightarrow \mu+1 \in \Sigma] \Rightarrow \Sigma = {}^{*}\mathbb{N}. \quad \Box$$

For the remainder of this section T is supposed to contain a theory of the ordered fields $\mathbb{Q}$ or $\mathbb{R}$, such that $T\langle\Omega\rangle$ contains a theory of Ω-rational or Ω-real numbers.

<u>Lemma 3:</u> (Special case of Cauchy's Principle) Let $\alpha(\mu)$ be an internal sequence of numbers such that $\alpha(\mu) > 0$ for finite μ. Then $\alpha(\mu) > 0$ for all $\mu \leqq \rho$ and some infinite ρ.

<u>Proof:</u> Let $a(k,m)$ be a double sequence of rational or real numbers such that $b(\Omega) = a(\Omega, m(\Omega)) = \alpha(\mu)$. Consider the following sequence of sets

$$S_n = \{m \mid a(n,m) \leqq 0\}$$

which can also be written

$$S_n = \{m(n) \mid a(n,m(n)) \leqq 0\}.$$

Thus, the following set is internal:

$$\Sigma = \{\mu \mid \alpha(\mu) \leqq 0\}.$$

For $\Sigma = \emptyset$ the proposition is true for any ρ. Otherwise Σ will contain a smallest element σ, by Lemma 1, which must be infinite. Let $\rho = \sigma-1$. $\Box$

<u>Lemma 4:</u> (Robinson's Lemma on Sequences) Let β_μ be an internal sequence such that $\beta_\mu \approx 0$ for all finite μ. Then there exists an infinite ρ such that $\beta_\mu \approx 0$ for $\mu \leqq \rho$.

Proof: Apply Lemma 3 to the sequence $\alpha_\mu = 1 - \mu|\beta_\mu|$. Π

For a finite Ω-real number ρ the standard part st ρ is the unique real number which is infinitesimally close to ρ. The monad mon σ of any Ω-number σ is the external set of all τ with $\tau \approx \sigma$. In the case of the field of Ω-rational numbers we consider the subring F of all finite numbers, including the set $I = \text{mon } 0$ of the infinitesimals. I is a maximal idea in F, and F/I is isomorphic to $\mathbb{R}$. In this case the standard part st is defined by the canonical homomorphism st: $F \to \mathbb{R}$. If $\lim r(n) = r_0$ in the sense of conventional analysis, then (fan) $|r(n) - r_0| < \varepsilon$ for every fixed rational $\varepsilon > 0$, hence $|\rho - r_0| < \varepsilon$ for $\rho = r(\Omega)$. In other words, $r_0 = \text{st } \rho$. Note that the converse need not be true: If $r_0 = \text{st } \rho$ then it may happen that $r(n)$ does not coverge at all. For example, $(-1)^\Omega$ will have standard part either $+1$ or -1, but $\lim(-1)^n$ does not exist. After these remarks we can state and prove:

Lemma 5: (Euler-Cauchy Condition for Convergence)
Let $\alpha(\mu)$ be a standard sequence. All $\alpha(\mu)$ for infinite μ belong to the same finite monad if and only if $\alpha(\mu) \approx \alpha(\lambda)$ for all infinite μ, λ. (In this case the real number corresponding to the monad is the limit of the sequence $\alpha(m)$ in the usual sense. Note that the essential part of the lemma is that the monad is finite, that is, consists of finite elements.)

Proof: If there is a monad to which all $\alpha(\mu)$ belong for infinite μ, then $\alpha(\mu) \approx \alpha(\lambda)$ for all infinite μ, λ. -
Let now be $\alpha(\mu) \approx \alpha(\Omega)$ for all infinite μ. If these numbers do not lie in a finite monad then $|\alpha(\Omega)|$ has to be infinite. Hence, for no finite M we can have (fan) $|a(n)| \leq M$. There exists a sequence $m_1 < m_2 < m_3 < \dots$ of natural numbers such that all $a(m_k)$ have the same sign, say, they are positive, and $a(m_{n+1}) \geq a(m_n) + 1$ for all n. By the Basic Definition we conclude $\alpha(\lambda) \geq \alpha(\mu) + 1$, where $\mu = m_\Omega$ and $\lambda = m_{\Omega+1}$. Since (m_n) is an increasing sequence of natural numbers both λ and μ are infinite, but $\alpha(\lambda) - \alpha(\mu) \geq 1$ contradicts $\alpha(\lambda) \approx \alpha(\mu)$. $\square$

A binary relation R is called concurrent on its left domain provided that for each finite set $x_1, \dots, x_n$ from the left domain there exists some y such that $R(x_k, y)$ holds simultaneously for all $k = 1, \dots, n$.

Lemma 6: (Weak Enlargement Property) Let R be a concurrent binary relation and $x_1, x_2, \dots$ a sequence of objects from T which are elements of the left domain of R. Then an object η of $T\langle\Omega\rangle$ exists such that $R(x_k,\eta)$ holds simultaneously for all $k = 1,2,\ldots$

Proof: Since (fan) $\bigvee_{y_n} \bigwedge_k \; k \leqq n \Rightarrow R(x_k,y_n)$ the following is true

$$\bigvee_{y_\Omega} \bigwedge_k \; k \leqq \Omega \Rightarrow R(x_k,y_\Omega). \text{ Let } \eta = y_\Omega. \;\square$$

The following stronger enlargement property will suffice for many applications:

Lemma 7: (Star-Finite Sets) Let E be any set of objects from T such that $|E| \leq |\mathbb{R}|$. Then there is an injective mapping of E into a segment $(1,2,3,\ldots,\mu)$ of the Ω-natural numbers. (The mapping will be external when E is infinite).

Proof: It will suffice to prove the proposition for some set of cardinality $|\mathbb{R}|$, say the standard real interval $I = \{r \in \mathbb{R} \mid 0 < r < 1\}$. For each $r \in I$ we take one of its dyadic representations,

$$r = \sum_{k=1}^{\infty} a_k(r)2^{-k}, \quad {}_k(r) = 0,1$$

and replace r by an Ω-rational number $\rho(r) \approx r$,

$$\rho(r) = \sum_{k=1}^{\Omega} a_k(r)2^{-k} .$$

Consider the Ω-natural number

$$\mu(r) = \sum_{k=1}^{\Omega} a_k(r)2^{+k} .$$

Then $\mu(r) \neq \mu(s)$ when $r \neq s$, since $\rho(r) \approx r$ and $\rho(s) \approx s$, thus $\rho(r) \neq \rho(s)$. Since $1 < \mu(r) \leq \sum_{k=1}^{\Omega} 2^k = 2^{\Omega+1} - 2$, μ is an injective mapping from the real interval to the segment $(1,2,3,\ldots,2^{\Omega+1} - 2)$. $\square$

4. An "Algebraization" of Real Analysis

A main feature of our approach is that the extension of a theory T to the larger theory $T\langle\Omega\rangle$ bears a striking analogy to the field extension procedure in algebra. In particular, starting from a theory T of rational numbers we obtain a theory of Ω-rational numbers which contains the reals in the shapes of monads of finite Ω-rational numbers.

It may be asked whether it is possible to find a quasi algebraic interpretation of the whole of real analysis in the same way. To this end, we consider the class P of Ω-polynomials with Ω-rational coefficients,

$$\phi(\xi) = \sum_{k=o}^{\mu} \alpha_k \xi^k ,$$

where $\mu = m(\Omega)$, $\alpha_k = a(\Omega,k(\Omega))$, $\xi = x(\Omega)$, and a sequence f_n of polynomials of finite degree with rational coefficients exists such that (fan) $f_n(x(n)) = \sum_{k=o}^{m(n)} a(n,k)\, x^k(n)$.

For polynomials the operations of differentiation and integration are defined in a purely algebraical way, and limiting processes are avoided. Actually the approach to real analysis via polynomials of infinite degree has its early and important roots in history. In particular, L. Euler used this approach most effectively.

A subclass L of P is of special interest. We shall say that $\phi \in P$ is of class L if

(i) $\phi(\xi)$ is finite for finite ξ,
(ii) $\phi(\xi) \approx \phi(\tilde{\xi})$ for finite $\xi \approx \tilde{\xi}$.

The latter property is sometimes called S-continuity; another very suggestive description is to say that ϕ is monad preserving, $\phi(\text{mon }\xi) \subseteq \text{mon }\phi(\xi)$ for finite ξ. As a consequence of (i) and (ii) we can define a real function f by $y = f(x)$ if $x = \text{mon }\xi$, ξ finite, and $y = \text{mon }\phi(\xi)$. We are going to prove that f is continuous in the usual sense of real analysis. Otherwise there would exist $a \in \mathbb{R}$, a rational $\varepsilon_o > 0$, and real a_m, $|a_m - a| < \frac{1}{m}$, such that $|f(a_m) - f(a)| > \varepsilon_o$ for finite $m = 1,2,3,\ldots$ Choose ξ_m and ξ Ω-rational and such that $a = \text{mon }\xi$, $a_m = \text{mon }\xi_m$. Then, from (ii), $|\xi_m - \xi| < \frac{1}{m}$ and $|\phi(\xi_m) - \phi(\xi)| > \varepsilon_o$ for finite m. We apply Lemma 3 (Cauchy's Principle) to the two sequences $s_m = \frac{1}{m} - |\xi_m - \xi|$ and $t_m = |\phi(\xi_m) - \phi(\xi)| - \varepsilon_o$.

It follows that for some infinite $m = \mu$ both $s_\mu > 0$ and $t_\mu > 0$. We obtain $\xi_\mu \approx \xi$, or $\text{mon }\xi_\mu = \text{mon }\xi = a$, but $|\phi(\xi_\mu) - \phi(\xi)| > \varepsilon_o$, hence $\text{mon }\phi(\xi_\mu) \neq \text{mon }\phi(\xi)$, a contradiction.

On the other hand, it is an easy consequence of the Weierstrass Approximation Theorem, applied to $f(x)$ for $|x| \leq \Omega$, that there exists

an Ω-polynomial ϕ such that $\phi(\xi) \approx f(\xi)$ for $|\xi| \leq \Omega$, in particular for all finite ξ. Thus we may conclude:
The continuous real functions $f: \mathbb{R} \to \mathbb{R}$ are represented by the classes of Ω-polynomials from L, where ϕ and ψ are equivalent if and only if $\phi(\xi) \approx \psi(\xi)$ for finite ξ.
The operation of integration, $\int_0^\xi \phi(\tau)d\tau$, maps L into L and preserves this equivalence relation. But this is not true for the operation of differentiation. For instance, the polynomials

$$\sigma(\xi) = \sum_{k=0}^{\Omega} \frac{(-1)^k}{(2k+1)!} \xi^{2k+1} \quad \text{and} \quad \gamma(\xi) = \sum_{k=0}^{\Omega} \frac{(-1)^k}{(2k)!} \xi^{2k}$$

represent the real functions sin and cos, and $\sigma'(\xi) = \gamma(\xi)$.
Hence $\phi(\xi) = \frac{\sigma(\Omega\xi)}{\Omega} \approx 0 = \psi(\xi)$, but $\phi'(\xi) = \gamma(\Omega\xi)$ which is a function not equivalent to 0. This leads in a natural way to a class $\mathcal{D}$ of Ω-polynomials, which is closed under the operations $+,-,\cdot$, differentiation, and integration: Φ is of class $\mathcal{D}$ if there is a $\phi \in L$ and a non-negative integer k such that $\Phi = \phi^{(k)}$.
The polynomials of class $\mathcal{D}$ represent, in a sense, derivatives of continuous functions. Thus it will be no surprise that they correspond to distributions. Actually, if we say that Φ, Ψ are equivalent in the distributional sense if and only if there exists ϕ, $\psi \in L$ and a non-negative integer k such that $\Phi = \phi^{(k)}$, $\Psi = \psi^{(k)}$ and $\phi(\xi) \approx \psi(\xi)$ for all finite ξ, then the equivalence classes are the distributions on $\mathbb{R}$. This is a well known result.

We have by now achieved an algebraization of parts of the conventional analysis including distributions. Continuous functions and distributions are represented by polynomials, and not only the rational operations but also differentiation and integration are algebraically defined for these polynomials.

5. Consistency of $T\langle\Omega\rangle$ as a Consequence of the Consistency of T

Suppose that the original theory T is consistent, and suppose that $T\langle\Omega\rangle$ contains a contradiction, that is, $C(\Omega) \equiv B(\Omega) \wedge \neg B(\Omega)$ is true in $T\langle\Omega\rangle$ for some admissible predicate B. To make this precise we have to specify the collection of theorems or true sentences which constitute $T\langle\Omega\rangle$. Among them will certainly be all those theorems which are obtained by the Basic Definition. Moreover, we shall admit as members of $T\langle\Omega\rangle$ all those sentences $D(\Omega)$ which are obtained from some $A(\Omega)$ of $T\langle\Omega\rangle$ by admissible deductions. In particular, a contradictory theorem $C(\Omega) = B(\Omega) \wedge \neg B(\Omega)$ would follow from some $A(\Omega)$, where (fan) $A(n)$. Since the same finite chain of conclusions which leads from $A(\Omega)$ to $C(\Omega)$, when applied to $A(n)$, would formally lead to $C(n)$, we obtain that (fan) $C(n)$, a contradiction in T itself. Thus, $T\langle\Omega\rangle$ is consistent if T is.

In view of the results of section 4 we have now a proof of the consistency of parts of real analysis including distributions, if T is a consistent elementary number theory of the rationals.

6. Models

How large is the theory $T\langle\Omega\rangle$? Is it equivalent to the theory of some reduced ultrapower $^*S$ when S is the basic set of the theory T? Surprisingly enough, the answer is yes.

Up to now, we had a rule for proving theorems in $T\langle\Omega\rangle$ which was given by the Basic Definition. By this rule, the theorem $(-1)^{\Omega} = +1 \vee (-1)^{\Omega} = -1$ could be proved. But we could neither conclude that $(-1)^{\Omega} = +1$ was true nor that $(-1)^{\Omega} = -1$ was true. A true formula like $A(\Omega) \vee B(\Omega)$ could be true though we where unable to decide whether $A(\Omega)$ was true or $B(\Omega)$ or both.

In the present section we shall adopt a different point of view: We shall interpret a true theorem of the form $A(\Omega) \vee B(\Omega)$ in the usual way, that means, we shall add a new rule of proving true theorems: If $A(\Omega) \vee B(\Omega)$ is proved by use of the Basic Definition, then at least one of the sentences $A(\Omega)$, $B(\Omega)$ will be said to be a true theorem. We shall add such a theorem to the collection $T\langle\Omega\rangle$ of true theorems. It may seem that a high degree of arbitraryness is creeping in. We shall show that our machinery provides for a unique model as soon as for a very restricted class of sentences of the type $A(\Omega) \vee B(\Omega)$ the decision will have been made whether $A(\Omega)$ or $B(\Omega)$ belongs to $T\langle\Omega\rangle$.

The new interpretation will enable us to represent $T\langle\Omega\rangle$ by set theoretical models. For each set $S \subseteq \mathbb{N}$ we consider its characteristic function, $\chi_S(n) = 1$ for $n \in S$, $\chi_S(n) = 0$ for $n \notin S$.

Since

(fan) $\chi_S(n) = 1 \vee \chi_S(n) = 0$,

we conclude that $\chi_S(\Omega) = 1 \vee \chi_S(\Omega) = 0$ is true in $T\langle\Omega\rangle$. In our new interpretation this means that actually either $\chi_S(\Omega) = 1$ is true or $\chi_S(\Omega) = 0$ is true. This leads to the definition of a set $U \subseteq P(\mathbb{N})$: $S \in U$ iff $\chi_S(\Omega) = 1$. We shall prove:

U *is a free ultrafilter on* $\mathbb{N}$.

(i) $\emptyset \notin U$ since (fan) $\chi_\emptyset(n) = 0$, hence $\chi_\emptyset(\Omega) = 0$.

(ii) Since

$$\chi_{S_1 \cup S_2}(\Omega) + \chi_{S_1 \cap S_2}(\Omega) = \chi_{S_1}(\Omega) + \chi_{S_2}(\Omega)$$

we conclude:

From $S_1, S_2 \in U$ follows $S_1 \cap S_2 \in U$.

(iii) If $S_o \in U$ and $S_1 \supseteq S_o$, then $S_1 \in U$. This is a consequence of

(fan) $\chi_{S_o}(n) = 1 \wedge S_1 \supseteq S_o \Rightarrow \chi_{S_1}(n) = 1$,

hence $\chi_{S_o}(\Omega) = 1 \wedge S_1 \supseteq S_o \Rightarrow \chi_{S_1}(\Omega) = 1$.

(iv) $\mathbb{N} \in U$, since $\chi_{\mathbb{N}}(\Omega) = 1$.

(v) For each $S \subseteq \mathbb{N}$, either $S \in U$ or $\bar{S} = \mathbb{N} \setminus S \in U$. This is a consequence of $\chi_S(\Omega) + \chi_{\bar{S}}(\Omega) = \chi_{\mathbb{N}}(\Omega) = 1$.

(vi) $\bigcap_{S \in U} S = \emptyset$. (Let $S_k = \mathbb{N} \setminus \{k\}$, then $\chi_{S_k}(\Omega) = 1$, hence $S_k \in U$, and $\bigcap_k S_k = \emptyset$.)

By (i), (ii), (iii), (iv) U is a filter, (v) is the ultrafilter property, and by (vi) U is free.

We proceed to consider predicates A(.). Let $\tau(B)$ denote the truth value of the sentence B, $\tau(B) = 1$ for B true and $\tau(B) = 0$ otherwise. The Basic Definition can be rewritten

If (fan) $\tau(A(n)) = 1$ then $\tau(A(\Omega)) = 1$.

Consider

$$S = S(A) = \{n \mid \tau(A(n)) = 1\}.$$

Then $\chi_S(n) = \tau(A(n))$, and since $\chi_S(\Omega) = \tau(A(\Omega)) = 1$, we obtain $S(A) \in U$. As a consequence of the ultrafilter property this yields finally:

$A(\Omega)$ *is true in the model of* $T\langle\Omega\rangle$ *if and only if* $\{n \mid A(n)$ is true in $T\} \in U$.

(Here U is the particular free ultrafilter on $\mathbb{N}$ which was obtained via the characteristic functions of subsets S of $\mathbb{N}$, and U depends on the choice that was made for the values of $\chi_S(\Omega)$.)

This shows that $T\langle\Omega\rangle$ is as large as any ultrapower model of the theory T with $\mathbb{N}$ as its index set.

Bibliography

[1] ROBINSON, A. and ZAKON, E.:
A set-theoretical characterization of enlargements. In "Applications of Model Theory", ed. by W.A.J. Luxemburg, New York 1969.

[2] SCHMIEDEN, C. and LAUGWITZ, D.:
Eine Erweiterung der Infinitesimalrechnung.
Math. Z. 69, 1-39 (1958).

[3] STROYAN, K.D. and LUXEMBURG, W.A.J.:
Introduction to the Theory of Infinitesimals. New York 1976.

Stochastic Integration in Hyperfinite Dimensional Linear Spaces

by

Tom L. Lindstrøm,
University of Oslo.[+])

Introduction.

The study of the nonstandard theory for stochastic integration was initiated by R.M. Anderson in [0]. Constructing a Brownian motion as the standard part of a hyperfinite random walk, Anderson showed how stochastic integrals with respect to the Brownian motion could be obtained from pathwise Stieltjes integrals with respect to the random walk. This intuitive definition of the stochastic integral as a Stieltjes integral - which it had proved impossible to obtain by standard methods - made it possible to give simplified proofs of the basic theorems of stochastic analysis. The theory has been used by H.J. Keisler to study stochastic differential equations [6], and it has also been extended to cover integration with respect to more general classes of martingales [8] - [10], [16] and [17].

While the papers mentioned above only treat stochastic integration with respect to martingales taking values in finite dimensional spaces, the theory of Hilbert space valued martingales has turned out to be of great importance in applications (see e.g. Chow [4], Kuo [7] and their references.) In this paper we shall develop the corresponding nonstandard theory - stochastic integration in hyperfinite dimensional linear spaces. We shall restrict our treatment to integration with respect to Brownian motions - in the classical theory this case involves some extra difficulties since the Brownian motion is not a process on the Hilbert space itself; we shall see how the nonstandard theory avoids these problems. The paper is divided into two parts; in the first part we obtain the basic results of the nonstandard theory and give an application to stochastic difference equations; in the second part we give the connection to the standard theory and develop the application to cover a class of partial stochastic differential equations.

Throughout the paper we shall work with polysaturated models of nonstandard analysis (see Stroyan and Luxemburg [15]); for an introduction to nonstandard measure and probability theory the reader should consult Loeb [12].

Part I.

1. Hyperfinite Dimensional Linear Spaces.

Let us define the notions we shall need about hyperfinite dimensional linear spaces and the mappings between them.

Let E be a nonstandard linear space with an inner product $\langle\cdot,\cdot\rangle$; then a countable, orthonormal set $\{v_n\}_{n\in\mathbb{N}}$ of elements from E is called a standard orthonormal set. An element $v \in E$ is called near-standard with respect to $\{v_n\}$ if for all $\epsilon \in \mathbb{R}_+$ there is an element u of the form $u = \sum_{n=1}^{k} a_n v_n$, $k \in \mathbb{N}$, $a_1,\ldots,a_n \in \mathbb{R}$, such that $\|v-u\| < \epsilon$. If E is hyperfinite dimensional, let $\{v_n\}_{n\leq\gamma}$ be an extension of $\{v_n\}_{n\in\mathbb{N}}$ to an orthonormal, internal basis for E; if $x = \Sigma x_n v_n$, then x is near-standard if and only if x_n is finite for $n \in \mathbb{N}$, and ${}^{o}\sum_{n=k}^{\gamma} x_n^2 \to 0$ as k tends to infinity in $\mathbb{N}$. In Part II we shall show how equivalence classes of the near-standard elements can be organized into a Hilbert space connected with E through a standard part map.

Let E and F be two hyperfinite dimensional linear spaces; a linear map $A : E \to F$ is called S-bounded if $\|A\|$ is finite.

Let $\{e_n\}$ and $\{f_n\}$ be standard orthonormal sets in E and F respectively; A is called <u>near</u>-<u>standard</u> (w.r.t. $\{e_n\}$ and $\{f_n\}$) if $A(e_n)$ is near-standard for all $n \in \mathbb{N}$. If A is S-bounded this means that A maps near-standard elements on near-standard elements, but this does not hold in general.

Let $\{e_n\}$ and $\{f_n\}$ be as above, and let $\{e_n\}_{n\leq\gamma}$ and $\{f_n\}_{n\leq\zeta}$ be extensions to internal orthonormal bases. A linear map $T : E \to F$ is called a <u>Hilbert</u>-<u>Schmidt</u> map if $\sum_{n=0}^{\infty} {}^{\circ}\|T^*f_n\|^2 = {}^{\circ}\sum_{n=0}^{\zeta}\|T^*f_n\|^2 < \infty$, where T^* is the adjoint of T. The sum $\sum_{n=0}^{\zeta}\|T^*f_n\|^2$ is easily seen to be independent of which orthonormal basis $\{f_n\}_{n\leq\zeta}$ we use, and we also have

$$\|T^*f_n\|^2 = \sum_{m=1}^{\gamma}\langle e_m, T^*f_n\rangle^2 = \sum_{m=1}^{\gamma}\langle Te_m, f_n\rangle^2$$

and hence

$$\sum_{n=1}^{\zeta}\|T^*f_n\| = \sum_{n=1}^{\zeta}\sum_{m=1}^{\gamma}\langle Te_m, f_n\rangle^2 = \sum_{m=1}^{\gamma}\sum_{n=1}^{\zeta}\langle Te_m, f_n\rangle^2 = \sum_{m=1}^{\gamma}\|Te_m\|^2 ,$$

which is the square of the usual Hilbert-Schmidt norm for T. We shall write $\|T\|_{(2)}$ for $(\sum_{n=1}^{\zeta}\|T^*f_n\|^2)^{\frac{1}{2}}$.

A Hilbert-Schmidt map is obviously S-bounded, and it is also near-standard:

$${}^{\circ}\sum_{m=k}^{\zeta}\langle Te_n, f_m\rangle^2 = {}^{\circ}\sum_{m=k}^{\zeta}\langle e_n, T^*f_m\rangle^2 \leq {}^{\circ}\sum_{m=k}^{\zeta}\|T^*f_m\|^2 \to 0$$

as $k \to \infty$ in $\mathbb{N}$.

2. Anderson Processes.

We shall now define "Brownian motions" on hyperfinite dimensional linear spaces. Let $\eta \in {}^*\mathbb{N} \setminus \mathbb{N}$, and consider the hyperfinite timeline $T = \{\frac{k}{\eta} : 0 \leq k \leq \eta\}$. When summing over elements of T we shall use the following convention: If $t = \frac{j}{\eta}$, we write

$$\sum_{j=o}^{t} A(s) \quad \text{for} \quad \sum_{k=o}^{j-1} A(\tfrac{k}{\eta});$$

hence $A(t)$ is not included in the sum $\sum_{j=o}^{t} A(s)$. We sometimes write Δt for $\frac{1}{\eta}$.

Let E be a hyperfinite dimensional linear space, and let $\{v_n\}_{n \leq \gamma}$ be any orthonormal, internal basis in E. Let $\Omega = (\{-1,1\}^{\gamma})^{\eta}$, and let P be the uniform probability measure on Ω, $P\{\omega\} = 2^{-\gamma \cdot \eta}$ for all $\omega \in \Omega$. We shall write ω_{js} for $\omega(k)(j)$, if $s = \frac{k}{\eta}$. Let $\chi : T \times \Omega \to E$ be the process defined by:

$$\chi(t,\omega) = \sum_{j=o}^{t} \sum_{j=1}^{\gamma} \frac{\omega_{js}}{\sqrt{\eta}} v_j.$$

We shall call such a process an Anderson process; it is obtained by having γ copies of Anderson's random walk running independently along orthonormal axes.

The variance of $\|\chi_t\|$ will be of magnitude $\gamma \cdot t$, and thus χ_t will not be near-standard for any orthonormal standard set. However, the next proposition tells us that χ has the right finite dimensional properties:

Proposition 1: Let $P : E \to E_o$ be the projection on the finite dimensional subspace E_o. Then ${}^o(P\chi_t)$ is a Brownian motion on E_o.

Proof: We first prove that $\|P\chi_t\|$ is finite a.e., using the independence of ω_{is} and ω_{jr} for $(i.s) \neq (j,r)$.

$$E(\|P\chi_t\|^2) = E\|\sum_{s=o}^{t} \sum_{j=1}^{\gamma} \frac{\omega_{js}}{\sqrt{\eta}} P(v_j)\|^2 = \sum_{s=o}^{t} \sum_{j=1}^{\gamma} \frac{1}{\eta}\|Pv_j\|^2 = t \sum_{j=1}^{\gamma} \|Pv_j\|^2 =$$

$$= t\|P\|^2_{(2)} = t \cdot \dim E_o < \infty.$$

To prove that ${}^oP\chi_t$ is a Brownian motion it is obviously enough to show that ${}^oP\chi_t - {}^oP\chi_s$ is gaussian distributed with mean zero and covariance matrix $(t-s)I$. Let us calculate the Fourier-transform:

For $z \in E_o$, we have:

$$\int e^{i\langle {}^oP\chi_t - {}^oP\chi_s, z\rangle} dL(P) = {}^o\!\int e^{i\langle P(\chi_t-\chi_s),z\rangle} dP = {}^o\!\int e^{i\langle \chi_t-\chi_s, z\rangle} dP$$

$$= {}^o\!\int e^{i \sum_{r=s}^{t} \sum_{j=1}^{\gamma} \frac{\omega_{jr}}{\sqrt{\eta}} \cdot z_j} dP = {}^o\Big(\prod_{r=s}^{t} \prod_{j=1}^{\gamma} \frac{e^{iz_j/\sqrt{\eta}} + e^{-iz_j/\sqrt{\eta}}}{2}\Big)$$

$$= {}^o\prod_{r=s}^{t} \prod_{j=1}^{\gamma} \cos\frac{z_j}{\sqrt{\eta}} = {}^oe^{\sum_{r=s}^{t} \sum_{j=1}^{\gamma} \ln\cos\frac{z_j}{\sqrt{\eta}}} \approx {}^oe^{-\sum_{r=s}^{t} \sum_{j=1}^{\gamma} \frac{z_j^2}{2\eta}} = e^{-\frac{(t-s)\langle z,z\rangle}{2}}$$

where we have used the second order Taylor formula for $\ln\cos$. This proves the proposition.

3. Stochastic Integration.

For each $t \in T$, define an equivalence relation $\sim_t$ on Ω by: $\omega \sim_t \omega'$ if and only if $\omega_{is} = \omega'_{is}$ for all i and all $s < t$. Let G_t be the internal algebra generated by $\sim_t$; then χ is a martingale with respect to $\langle \Omega, \{G_t\}, P\rangle$. A process $X: T \times \Omega \to K$ - where K is hyperfinite dimensional linear space - is said to be adapted to $\langle\Omega,\{G_t\},P\rangle$ if each X_t is G_t-measurable.

If E and F are two hyperfinite dimensional linear spaces, L(E,F) denotes the linear maps from E to F. We now define the integrands we shall work with in this paper. Let $\{f_n\}_{n\in\mathbb{N}}$ be a standard orthonormal set in F:

Definition 2: A process $X: T\times\Omega \to L(E,F)$ is said to be in $m^2(E,F)$ if it is adapted to $\langle\Omega,\{G_t\},P\rangle$ and

$$\lim_{k\to\infty} E\{\int_0^1 {}^{o}\sum_{m=1}^{k}\|X(s,\omega)^*(f_m)\|^2 ds\} = {}^{o}E\{\int_0^1 \|X(s,\omega)\|_{(2)}^2 ds\} < \infty.$$

It follows directly from Definition 2 that if $X\in m^2(E,F)$ then $X(s,\omega)$ is Hilbert-Schmidt for almost all (s,ω); the placing of the standard parts also ensures us that $\|X(s,\omega)\|_{(2)}^2$ is S-integrable in the product measure.

Definition 3: If $\chi: T\times\Omega \to E$ is an Anderson process and $X\in m^2(E,F)$, we define the stochastic integral $\int X d\chi$ as the process from $T\times\Omega$ into F defined by

$$(\int X d\chi)(t,\omega) = \sum_{s=0}^{t} X(s,\omega)(\Delta\chi(s,\omega)).$$

Here $\Delta\chi(s,\omega) = \chi(s+\frac{1}{\eta},\omega) - \chi(s,\omega)$.

Since X is adapted, $\int X d\chi$ is a martingale with respect to $\langle\Omega,\{G_t\},P\rangle$.

In our next theorem we shall prove some regularity properties of the stochastic integral. An internal function $f: T\to F$ is said to be S-continuous if $\|f(t)-f(s)\|\approx 0$ each time $t\approx s$; a process $X: T\times\Omega \to F$ is S-continuous if almost all paths are S-continuous. A process is near-standard if for almost all ω,

$X(t,\omega)$ is near-standard for all $t \in T$.

If $M : T \times \Omega \to {}^*\mathbb{R}$ is a martingale such that $E(M_t^2) < \infty$ for all $t \in T$, let the <u>quadratic variation</u> $[M]$ be the process defined by $[M](t) = \sum_{s=0}^{t} \Delta M(s)^2$. Then by Theorem 14 of [8], M is S-continuous if and only if $[M]$ is. We may now prove:

<u>Theorem 4</u>: Let $\chi : T \times \Omega \to E$ be the Anderson process, and let $X \in m^2(E,F)$. Then $Y = \int X d\chi$ is near-standard and S-continuous.

Proof: We have

$$E(Y_k(t)^2) = E((\sum_{s=0}^{t} \langle X_s(\Delta\chi_s), f_k\rangle)^2) = E(\sum_{s=0}^{t} \sum_{j=1}^{\gamma} \frac{\omega_{js}}{\sqrt{\eta}} \langle X_s(v_j), f_k\rangle)^2 =$$

$$= E(\sum_{s=0}^{t} \sum_{j=1}^{\gamma} \frac{1}{\eta} \langle X_s(v_j), f_k\rangle^2) = E(\int_0^t \sum_{j=1}^{\gamma} \langle v_j, X_s^*(f_k)\rangle^2 dt) = E(\int_0^t \|X_s^*(f_k)\|^2 dt)$$

where we have used the independence of the ω_{js}. By Doob's inequality

$$0 \le E(\sup \sum_{k=m}^{\zeta} Y_k(t)^2) \le 4E(\int_0^t \sum_{k=m}^{\zeta} \|X(s,\omega)^*(f_k)\|^2 dt) \to 0$$

as $m \to \infty$ in $\mathbb{N}$, by the definition of $m^2(E,F)$. Hence Y is near-standard a.e.

Let us now prove that Y is S-continuous; we do this by first proving that each component Y_k is S-continuous. We shall apply the result from [8] mentioned above, but we first perform the following trick: Divide each interval $[t, t+\frac{1}{\eta}\rangle$ of the time-line into γ points, and construct a new martingale where the jump $\frac{\omega_{js}}{\sqrt{\eta}}\langle X_s(v_j), f_k\rangle$ appears at time $s + \frac{j-1}{\eta\gamma}$. This gives us a new martingale $\tilde{Y}_k : T' \times \Omega \to {}^*\mathbb{R}$ which agrees with Y_k on the points of the old time-line. $\tilde{Y}_k$ has the following quadratic variation:

$$[\tilde{Y}_k](t) = \sum_{s=0}^{t} \sum_{j=1}^{\gamma} \frac{1}{\eta}\langle X_s(v_j), f_k\rangle^2 = \int_0^t \|X_s^*(f_k)\|^2 ds.$$

Since $\|X(s,\omega)^*(f_k)\|^2 \leq \|X(s,\omega)\|^2_{(2)}$ is S-integrable in the product measure, $s \to \|X(s,\omega)^*(f_k)\|^2$ is S-integrable for almost all ω, and hence $[\tilde{Y}_k]$ is S-continuous for all those ω. Hence $\tilde{Y}_k$- and also Y_k- are S-continuous.

It is now easy to prove that Y is S-continuous. If $s \approx t, n \in N$, then $\sum_{k=1}^{n}(Y_k(t) - Y_k(s))^2 \approx 0$ a.e.; by Robinson's Sequential Lemma there is a $\xi \in {}^*\mathbb{N} \setminus \mathbb{N}$ such that $\sum_{k=1}^{\xi}(Y_k(t) - Y_k(s))^2 \approx 0$. But since Y is near-standard $\sum_{k=\xi+1}^{\zeta}(Y_k(t) - Y_k(s))^2 \approx 0$. Hence $\sum_{k=1}^{\zeta}(Y_k(t) - Y_k(s))^2 \approx 0$, and the theorem is proved.

4. Stochastic Difference Equations with Unbounded Operators.

In this section we shall give an application to the theory of stochastic difference equations. We shall study equations of the form

$$x(t) = x_o + \sum_{s=0}^{t} A(s,x(s))\Delta t + \sum_{s=0}^{t} \xi(s,x(s))\Delta\chi(s),$$

where χ is an Anderson process on E, and $A: T \times F \times \Omega \to F$ and $\xi: T \times F \to L(E,F)$ are suitable mappings. This equation obviously has a unique solution; the question is rather whether that solution is near-standard and S-continuous. If the operators $A_{t,\omega}: F \to F$ are uniformly S-Lipschitz continuous (not necessarily linear), then it is not difficult to find conditions on A, ξ and x_o ensuring that the solution has these properties; since the methods are similar to those we shall employ below, we do not consider this problem in greater detail. Instead we study the case where A is linear, S-unbounded, but independent of t and ω; i.e. an equation of the form

$$(*) \quad x(t) = x_o + \sum_{s=o}^{t} A(x(s))\Delta t + \sum_{s=o}^{t} \xi(s,x(s))\Delta\chi(s),$$

where $A : F \to F$ is linear, but not necessarily S-bounded.

By the <u>semi-group</u> $\{T_t\}_{t\in T}$ <u>generated by</u> A, we mean the operators $T_t = (I + \frac{A}{\eta})^{t\eta}$. Obviously, $T_o = I$, $T_{t+s} = T_t \cdot T_s$ and $\frac{T_{\Delta t} - I}{\Delta t} = A$. A is called the <u>infinitesimal generator</u> of $\{T_t\}$. The semi-group is called <u>strongly continuous</u> if there is an $M \in \mathbb{N}$ such that $\|T_t\| \leq M$ for all $t \in T$, and $t \to T_t(v)$ is S-continuous for all near-standard $v \in F$. As in the standard theory, an S-unbounded operator A may well be the infinitesimal generator of a strongly continuous semi-group $\{T_t\}$; in fact the Hille-Yosida-Phillips theorem (see e.g. Reed and Simon [14]) carries over to the nonstandard theory. One may also prove that if $\|A\| \ll \eta$, then if A is near-standard so are all T_t in a strongly continuous semi-group generated by A.

By induction it is trivial to see that (*) is equivalent to

$$(**) \quad x(t) = T_t x_o + \sum_{o}^{t} T_{t-\Delta t-s}\, \xi((s,x(s))\Delta\chi(s)$$

Hence we can transform difference equations governed by unbounded operators A, into difference equations governed by bounded operators T_t. We shall use this to prove the regularity results we want about the solutions of (*). First we prove the following simple estimates:

Lemma 5: Let $\delta, L > 0$ and assume that $z : T \to {}^*\mathbb{R}$ satisfies

$$z(t) \leq \delta + L \sum_{s=0}^{t} z(s)\Delta t \quad \text{for all} \quad t \in T.$$

Then

$$z(t) \leq \delta e^{Lt} \quad \text{and} \quad \sum_{0}^{1} z(t)\Delta t \leq \frac{\delta}{L}(e^{L} - 1) \quad \text{for all} \quad t \in T.$$

Proof: We prove the first inequality by induction on t: It is obvious for $t = 0$, and we have:

$$z(t+\Delta t) \leq \delta + L \sum_{s=0}^{t+\Delta t} z(s)\Delta t \leq \delta + L \sum_{s=0}^{t+\Delta t} \delta e^{Ls} \Delta t = \delta(1 + L\Delta t \sum_{s=0}^{t+\Delta t} e^{Ls})$$

$$= \delta(1 + L\Delta t \frac{e^{L(t+\Delta t)} - 1}{e^{L\Delta t} - 1}) \leq \delta(1 + L\Delta t \frac{e^{L(t+\Delta t)} - 1}{L\Delta t}) \leq \delta e^{L(t+\Delta t)},$$

using the induction hypothesis, the formula for the sum of a geometric series, and the inequality $e^{L\Delta t} \geq 1 + L\Delta t$.

The second inequality follows easily from the first.

We may now prove our result:

Theorem 6: Let $A : F \to F$ be a linear, near-standard operator generating a strongly continuous semi-group $\{T_t\}$ of near-standard operators, and let $\chi : T \times \Omega \to E$ be an Anderson process. Let $\xi : T \times F \to L(E,F)$ be such that there is a $K \in \mathbb{N}$ with $\|\xi(t,u) - \xi(t,v)\|_{(2)} \leq K\|u-v\|$ for all t,u,v; that for almost all t, $\xi(t,u)$ is Hilbert-Schmidt for all near-standard u; and that $t \to \|\xi(t,0)\|^2_{(2)}$ is S-integrable. Finally, let x_o be a G_o-measurable initial condition with $\|x_o\|^2$ S-integrable. Then

$$x(t) = x_o + \sum_{s=0}^{t} A(x(s))\Delta t + \sum_{s=0}^{t} \xi(s,x(s))\Delta\chi(s)$$

is near-standard and S-continuous.

Proof: We consider the equivalent equation

$$x(t) = T_t x_o + \sum_{s=0}^{t} T_{t-\Delta t-s}\xi(s,x(s))\Delta\chi(s).$$

Let us first prove that if τ is a stopping time, then $E\|x(t\wedge\tau)\|^2 < \infty$ for all $t \in T$: If $M \in \mathbb{N}$ is such that $\|T_t\| \leq M$ for all $t \in T$, we have:

$$E\|x(t\wedge\tau)\|^2 = E\|T_{t\wedge\tau}x_o + \sum_{s=0}^{t\wedge\tau} T_{t\wedge\tau-\Delta t-s}\xi(s,x(s\wedge\tau))\Delta\chi(s)\|^2$$

$$= E\|T_{t\wedge\tau}x_o\|^2 + E\|\sum_{s=0}^{t\wedge\tau} T_{t\wedge\tau-\Delta t-s}\xi(s,x(s\wedge\tau))\Delta\chi(s)\|^2$$

$$\leq M^2E\|x_o\|^2 + E\sum_{s=0}^{t\wedge\tau}\|T_{t\wedge\tau-\Delta t-s}\xi(s,x(s\wedge\tau))\|^2_{(2)}\Delta t$$

$$\leq M^2E\|x_o\|^2 + M^2E\sum_{s=0}^{t\wedge\tau}\|\xi(s,x(s\wedge\tau))\|^2_{(2)}\Delta t$$

$$\leq M^2(E\|x_o\|^2 + 2E\sum_{s=0}^{t\wedge\tau}\|\xi(s,0)\|^2_{(2)}\Delta t + 2E\sum_{s=0}^{t\wedge\tau}\|\xi(s,x(s)) - \xi(s,0)\|^2_{(2)}\Delta t)$$

$$\leq M^2(E\|x_o\|^2 + 2E\sum_{s=0}^{1}\|\xi(s,0)\|^2_{(2)}\Delta t + 2K^2\sum_{s=0}^{t}E\|x(s\wedge\tau)\|^2\Delta t),$$

where we have used that x_o is G_o-measurable, that stochastic integrals are martingales, and the trivial inequality $(a+b)^2 \leq 2a^2 + 2b^2$. Using Lemma 5 with $\delta = M^2(E\|x_o\|^2 + 2E\sum_{s=0}^{1}\|\xi(s,0)\|^2_{(2)}\Delta t)$ and $L = 2K^2M^2$, we see that $E\|x(t\wedge\tau)\|^2 < \infty$. It follows that a.e. $\sup_{s\in T}\|x(s)\| < \infty$; for if not there exists a $\gamma \in {}^*\mathbb{N}\setminus\mathbb{N}$ such that $\sup_{s\in T}\|x(s)\| > \gamma$ on a set of non-infinitesimal measure, and then ${}^oE\|x(t\wedge\tau_\gamma)\|^2 = \infty$, where $\tau_\gamma = \min\{t\in T : \|x(t)\| \geq \gamma\}$.

We now prove that x is near-standard: Let $\{f_n\}_{n\in\mathbb{N}}$ be the standard orthonormal set in F, and let $\{f_n\}_{n\leq\zeta}$ be an extension to an orthonormal basis in F. Let P_m be the projection on the linear span of $\{f_{m+1},\ldots,f_\zeta\}$, and $P_m^\perp$ the projection on the linear span of $\{f_1,\ldots,f_m\}$. For stopping times τ, we have by calculations similar to the one above:

$$E\|P_m x(t\wedge\tau)\|^2 \leq E\|P_m T_{t\wedge\tau} x_o\|^2 + E\sum_{s=o}^{t\wedge\tau} \|P_m T_{t\wedge\tau-\Delta t-s}\xi(s,x(s\wedge\tau))\|^2_{(2)}\Delta t$$

$$\leq E\|P_m T_{t\wedge\tau} x_o\|^2 + E\sum_{s=o}^{t\wedge\tau} \|P_m T_{t\wedge\tau-\Delta t-s}\xi(s,P_n x(s\wedge\tau)+P_n^\perp x(s\wedge\tau))\|^2_{(2)}\Delta t.$$

But now

$$\|P_m T_{t\wedge\tau-\Delta t-s}\xi(s,P_n x(s\wedge\tau)+P_n^\perp x(s\wedge\tau))\|_{(2)}$$

$$\leq \|P_m T_{t\wedge\tau-\Delta t-s}\xi(s,P_n x(s\wedge\tau)+P_n^\perp x(s\wedge\tau)) - P_m T_{t\wedge\tau-\Delta t-s}\xi(s,P_n^\perp x(s\wedge\tau))\|_{(2)}$$

$$+ \|P_m T_{t\wedge\tau-\Delta t-s}\xi(s,P_n^\perp x(s\wedge\tau))\|_{(2)}$$

$$\leq MK\|P_n x(s\wedge\tau)\| + \|P_m T_{t\wedge\tau-\Delta t-s}\xi(s,P_n^\perp x(s\wedge\tau))\|_{(2)}$$

and hence

$$E\|P_m x(t\wedge\tau)\|^2 \leq E\|P_m T_{t\wedge\tau} x_o\|^2 + 2M^2K^2 E\sum_{s=o}^{t\wedge\tau} \|P_n x(s\wedge\tau)\|^2\Delta t$$

$$+ 2E\sum_{s=o}^{t\wedge\tau} \|P_m T_{t\wedge\tau-\Delta t-s}\xi(s,P_n^\perp x(s\wedge\tau))\|^2_{(2)}$$

$$\leq E\|P_m T_{t\wedge\tau} x_o\|^2 + 2M^2K^2 E\sum_{s=o}^{t\wedge\tau} \|P_m x(s\wedge\tau)\|^2\Delta t$$

$$+ 2M^2K^2 E\sum_{s=o}^{t\wedge\tau} |\,\|P_m x(s\wedge\tau)\|^2 - \|P_n x(s\wedge\tau)\|^2|\,\Delta t$$

$$+ 2E\sum_{s=o}^{t\wedge\tau} \|P_m T_{t\wedge\tau-\Delta t-s}\xi(s,P_n^\perp x(s\wedge\tau))\|^2_{(2)}\Delta t\,.$$

For each $k\in\mathbb{N}$, let τ_k be the stopping time

$$\tau_k = \min\{t\in T : \|x(t)\| \geq k\}.$$

If σ is a stopping time, let σ_k denote $\sigma \wedge \tau_k$. We now apply the inequality above with $\tau = \sigma \wedge \tau_k$:

Let $\varepsilon > 0$, $\varepsilon \in \mathbb{R}$, be given. Choose $n \in \mathbb{N}$ so large that

$$\sum_{s=0}^{1\wedge\sigma_k} E\left|\|P_n x(s\wedge\sigma_k)\|^2 - \|P_m x(s\wedge\sigma_k)\|^2\right| \Delta t < \varepsilon$$

for all $m > n$; this is possible since $E \sum_{s=0}^{1\wedge\sigma_k} \|P_n x(s\wedge\sigma_k)\|^2 \Delta t$ is S-bounded and decreasing. Having chosen such an n, the sequence $^{o}\|P_m T_{t\wedge\sigma_k - \Delta t - s}\xi(s, P_n^{\perp} x(s\wedge\sigma_k))\|_{(2)}^2$ tends to zero as m goes to infinity in $\mathbb{N}$ a.e. in $T \times \Omega$, since $T_{t\wedge\sigma_k - \Delta t - s}$ is near-standard, and $\xi(s, P_n^{\perp} x(s\wedge\sigma_k))$ is Hilbert-Schmidt. Also $\|P_m T_{t\wedge\sigma_k - \Delta t - s}\xi(s, P_n^{\perp} x(s\wedge\sigma_k))\|_{(2)}^2$ is S-integrable since $\|\xi(s,0)\|_{(2)}^2$ is, and $\|P_n^{\perp} x(s\wedge\sigma_k)\|$ is S-bounded. Hence we may choose m such that $E \sum_{s=0}^{1\wedge\sigma_k} \|P_m T_{t\wedge\sigma_k - \Delta t - s}\, \xi(s, P_n^{\perp} x(s\wedge\sigma_k))\|_{(2)}^2 \Delta t < \varepsilon$; and since x_0 is near-standard and $T_{t\wedge\sigma_k}$ is near-standard and S-bounded, we may choose it such that $E\|P_m T_{t\wedge\sigma_k} x_0\|^2 < \varepsilon$ as well.

The inequality now becomes

$$E\|P_m x(t\wedge\sigma_k)\|^2 \leq \varepsilon + 2M^2K^2\varepsilon + 2\varepsilon + 2M^2K^2 E \sum_{s=0}^{t\wedge\sigma_k} \|P_m x(s\wedge\sigma_k)\|^2 \Delta t.$$

Applying Lemma 5 with $\delta = \varepsilon(2M^2K^2 + 3)$ and $L = 2M^2K^2$, we see that we can get $^{o}E\|P_m x(t\wedge\sigma_k)\|^2$ as small as we wish by choosing m large enough. Since by the first part of the proof $\tau_k \to 1$ a.e. as $k \to \infty$, this implies that x is near-standard.

We finally prove that x is S-continuous: Since x_0 is near-standard and $\{T_t\}$ is a strongly continuous semi-group of near-standard operators, $t \to T_t x_0$ is S-continuous and near-standard. Also, $t \to \sum_{s=0}^{t} T_{t-\Delta t-s}\xi(s,x(s))\Delta\chi(s)$ is near-standard. Now

$$\sum_{s=0}^{t} T_{t-\Delta t-s}\xi(s,x(s))\Delta\chi(s) - \sum_{s=0}^{r} T_{r-\Delta t-s}\xi(s,x(s))\Delta\chi(s)$$

$$= \sum_{s=r}^{t} T_{t-\Delta t-s}\xi(s,x(s))\Delta\chi(s) - (T_{t-r}-I)\sum_{o}^{r} T_{r-s-\Delta t}\xi(s,x(s)\Delta\chi(s)$$

Since $\sum_{o}^{r} T_{r-\Delta t-s}\xi(s,x(s))\Delta\chi(s)$ is near-standard and $\{T_t\}$ is strongly continuous, the last term is infinitesimal for $t\approx r$.

Turning to the first term, we consider the martingales

$$r \to \sum_{s=0}^{r} T_{t-\Delta t-s}\xi(s,x(s))\Delta\chi(s) = T_{t-r}\sum_{s=0}^{r} T_{r-\Delta t-s}\xi(s,x(s))\Delta\chi(s).$$

By what we have proved above and the near-standardness of T_{t-r}, these martingales are near-standard a.e. Since $s \to \|\xi(s,0)\|^2_{(2)}$ is S-integrable, so is $\|T_{t-s-\Delta t}\xi(s,x(s))\|^2_{(2)}$ for almost all ω, and hence also $\|T_{t-\Delta t-s}\xi(s,x(s))^*(f_k)\|^2$. But by the proof of Theorem 4 this is all we need for proving that the martingale is S-continuous a.e. Let $\{q_n\}_{n\in\mathbb{N}}$ be a S-dense subset of T; then there is a set of measure one where all the martingales

$$r \to \sum_{s=0}^{r} T_{q_n-\Delta t-s}\xi(s,x(s))\Delta\chi(s)$$

are S-continuous. But since

$$T_{q_n-s}\sum_{s=r}^{t} T_{t-s}\xi(s,x(s))\Delta\chi(s) = \sum_{s=r}^{t} T_{q_n-s-\Delta t}\xi(s,x(s))\Delta\chi(s)$$

it follows from the S-density of $\{q_n\}$ and the strong continuity of T_{q_n-s}, that $\sum_{s=r}^{t} T_{t-s-\Delta t}\xi(s,x(s))\Delta\chi(s)$ must be infinitesimal on this set for $t\approx r$. Hence x is S-continuous, and the theorem is proved.

Part II

We shall now give the connection between the nonstandard theory developed in the first part of this paper and the standard theory for stochastic integration in Hilbert spaces. Combined with Theorem I-6 this will give us an existence result for solutions of a class of partial stochastic differential equations.

1. Standard parts.

Let E be a hyperfinite dimensional linear space with an inner product $\langle\cdot,\cdot\rangle$, and let $|\cdot|$ be a norm on E (not necessarily generated by $\langle\cdot,\cdot\rangle$). Assume that $\{v_n\}_{n\in\mathbb{N}}$ is a standard orthonormal set in E, and that $0<{}^o|v_n|<\infty$ for all $n\in\mathbb{N}$. An element $v\in E$ is called $|\cdot|$-near-standard if for all $\varepsilon\in\mathbb{R}_+$, there is an element $u\in E$ of the form $u=\sum_{n=1}^{k}a_nv_n$, $k\in\mathbb{N}$, $a_1,\ldots,a_n\in\mathbb{R}$, such that $|v-u|<\varepsilon$. The set of $|\cdot|$-near-standard elements of E is denoted by $Ns_{|\cdot|}(E)$; this set depends heavily on the norm $|\cdot|$. Define an equivalence relation $\sim_{|\cdot|}$ on E by $x\sim_{|\cdot|}y \iff |x-y|\approx 0$, and let ${}^oE_{|\cdot|}$ denote the set $Ns_{|\cdot|}(E)/\sim_{|\cdot|}$. Let ${}^o|\cdot|: {}^oE_{|\cdot|}\to\mathbb{R}$ be defined by ${}^o|{}^ox|=st|x|$, where ${}^ox$ denotes the equivalence class of $x\in Ns_{|\cdot|}(E)$. We then have:

<u>Lemma 1</u>: $({}^oE_{|\cdot|},{}^o|\cdot|)$ is a Banach space, and the linear span of $\{{}^ov_n\}$ is dense in ${}^oE_{|\cdot|}$.

Proof: $({}^{o}E_{|\cdot|}, {}^{o}|\cdot|)$ is obviously a normed space, and the linear span of $\{{}^{o}v_n\}$ is dense by definition; hence we only have to show that ${}^{o}E_{|\cdot|}$ is complete: Let $\{{}^{o}x_n\}_{n\in\mathbb{N}}$ be a Cauchy-sequence in ${}^{o}E_{|\cdot|}$, and let $\{x_n\}_{n\in{}^*\mathbb{N}}$ be an extension of $\{x_n\}_{n\in\mathbb{N}}$ to an internal sequence such that $|x_n - x_m| \approx 0$ for all $n,m \in {}^*\mathbb{N} \setminus \mathbb{N}$; then since all $x_n, n \in \mathbb{N}$, are near-standard, so is x_γ, and $\{{}^{o}x_n\}$ converges to ${}^{o}x_\gamma$. This proves the lemma.

The lemma shows us how to turn nonstandard linear spaces into standard ones; and if $|\cdot|$ is the norm generated by $\langle\cdot,\cdot\rangle$, then $({}^{o}E_{|\cdot|}, {}^{o}|\cdot|)$ is a Hilbert space with $\{{}^{o}v_n\}$ as an orthonormal basis, and the map $x \to {}^{o}x$ maps the near-standard elements of E to elements in the Hilbert space.

The mapping $st_{|\cdot|} : Ns_{|\cdot|}(E) \to {}^{o}E_{|\cdot|}$ defined by

$$st_{|\cdot|}(x) = {}^{o}x$$

will be called a <u>standard part map</u> eventhough it is not exactly what is usually called by that name. However, it has most of the properties of ordinary standard parts, for instance:

<u>Lemma 2</u>: Let E, $|\cdot|$ and $\{v_n\}$ be as above. Let $\langle \Omega, G, P\rangle$ be a nonstandard probability space, and let $Y : \Omega \to {}^{o}E_{|\cdot|}$ be an $L(G)$-measurable random variable. Then there exists an internal, G-mesurable random variable $Z : \Omega \to E$ such that ${}^{o}Z(\omega) = Y(\omega)$ for $L(P)$-a.a. $\omega \in \Omega$.

Proof: Exactly as the proof of Theorem III. 4.3. in Anderson [1].

In the lemma above, Z is called a <u>lifting</u> of Y.

2. Brownian Motion in Hilbert Spaces.

To define Brownian motions in a Hilbert space H, we must first define Gaussian measures on H. A natural definition would be a measure on H such that the projection on any finite dimensional subspace E of dimension n, is a Gaussian measure with mean zero and variance nI; but it is easy to see that such a measure do not exist; the mass disappears at infinity. Instead we must put the measure on a larger space, proceeding in the following way:

If H is a real, separable Hilbert space, let I be the set of finite dimensional subspaces of H; and if $E,F \in I$, $E \subset F$, let $P_{E,F}: F \to E$ be the projection. A <u>normally distributed cylindrical measure</u> on H is a net $\{\mu_E\}_{E \in I}$, where each μ_E is a Gaussian measure on E with mean zero and covariance $(\dim E)I$; and if $E \subset F$, then $\mu_E = P_{E,F}(\mu_F)$. The measures $\{\mu_E\}$ clearly induces a finitely additive measure μ on H. If $|\cdot|$ is a norm on H (not necessarily the Hilbert-space norm $\|\cdot\|$) $|\cdot|$ is said to be <u>measurable</u> if for all $\varepsilon > 0$, there exists a finite dimensional projection P_o such that

$$\mu\{x : |Px| > \varepsilon\} < \varepsilon$$

for all finite dimensional projections $P \perp P_o$. If H is infinite dimensional, then the Hilbert space norm $\|\cdot\|$ is <u>not</u> measurable.

Let B be the completion of H in $|\cdot|$; we transport the cylindrical measure μ on H to a cylindrical measure ν on B by defining for all $y_1, \ldots\ldots, y_n$ in the dual B^* of B:

$$\nu\{x \in B : (y_1(x), \ldots, y_n(x)) \in E\} = \mu\{x \in H : (\langle x,y_1\rangle, \ldots, \langle x,y_n\rangle) \in E\}$$

for all Borel sets E in $\mathbb{R}^n$. (Here we have identified B^* with a subset of H.)

We now have the following result due to L. Gross [5]; for a standard proof see also Kuo [7]:

Theorem 3: If $|\cdot|$ is measurable, the ν has an extension to a countably additive Borel-measure on B.

A simple nonstandard proof of this theorem was given in [11]. The idea is as follows: Let $\{v_n\}$ be an orthonormal basis for H, and let E be a hyperfinite dimensional subspace of ${}^*H$ generated by an initial segment of ${}^*(\{v_n\}_{n\in\mathbb{N}})$. By the measurability of $|\cdot|, L(\tilde{\mu}_E)$ is $|\cdot|$-near-standard on E, and we may define a measure $\tilde{\nu}$ on B by $\tilde{\nu} = st_{|\ |}(L(\tilde{\mu}_E))$. It is not difficult to see that this is a countably additive Borel-measure extending ν.

If H and B are as above and $i : H \to B$ is the inclusion map, the triple (i,H,B) is called an abstract Wiener-space. The measure of Theorem 3 is called the Wiener-measure on (i,H,B); and by the Wiener-measure of variance parameter t, we shall mean the measure obtained in the same way, only starting with a cylindrical measure $\{\mu_E\}$ of Gaussian measures with covariance $t \dim E \cdot I$.

By a Brownian motion on (i,H,B) we shall mean a process $W : [0,1] \times \Omega \to B$ with independent increments, and such that when $s < t$, $W(t) - W(s)$ induces a Wiener-measure of variance parameter $t-s$ on B.

Let now E be a hyperfinite dimensional subspace of ${}^*H$ containing H, and χ be an Anderson process on E (recall the definition from I, section 2). If $|\cdot|$ is a measurable norm, it follows from the nonstandard proof of Theorem 3 that χ is

$|\cdot|$-near-standard a.e., and by Lemma 1 and Proposition I-1 it follows that $st_{|\ |}(\chi)$ is a Brownian motion on (i,H,B).

3. The Nonstandard Approach to Stochastic Integration.

Let $\chi : T \times \Omega \to E$ be an Anderson process, where E is as in the last section. Let $\{G_t\}_{t \in T}$ be the stochastic basis for χ as in section I-4, and define for $t \in [0,1]$

$$F_t = \bigcup_{s \approx t} \sigma(L(G_s) \cup N)$$

where N are the null-sets of Ω. As in [9] one proves that $\{F_t\}$ is an increasing family of σ-algebras, and that $W = {}^o\chi$ is a martingale with respect to the basis $\langle \Omega, \{F_t\}, L(P) \rangle$.

A process $\xi : [0,1] \times \Omega \to V$ taking values in some Banach space V is said to be <u>progressively measurable</u> if ξ is product measurable and ξ_t is F_t-measurable for all t. Let $\xi \in M[V]$ if $E\int_0^s \|\xi(t)\|^2 dt < \infty$, and ξ is progressively measurable. We shall define stochastic integrals of the form $\int \xi dW$ when $\xi \in M[L_{(2)}(H,K)]$, where $L_{(2)}(H,K)$ are the Hilbert-Schmidt operators from H into another Hilbert space K.

We first show how to replace an integrand $\xi \in M[L_{(2)}(H,K)]$ by an integrand $X \in m^2(E,F)$, where F is a hyperfinite dimensional subspace of ${}^*K$, $K \subset F$:

Let $\{e_m\}$ be an orthonormal basis in H, and extend it to an orthonormal internal basis $\{\tilde{e}_n\}_{n \le \gamma}$ for E. If $T : H \to K$ is Hilbert-Schmidt, let $\tilde{T} : E \to F$ be $P_F{}^*T \restriction E$; $\tilde{T}$ is then obviously Hilbert-Schmidt. Choose an orthonormal basis $\{T_n\}$ for $L_{(2)}(H,K)$, and let $\{\tilde{T}_n\}$ be an orthonormal standard set in $L(E,F)$. If $\xi \in M[L_{(2)}(H,K)]$, we get from Lemma 2 and (the proof of) Theorem 7.4 in Keisler [6] that ξ has an adapted lifting Y taking values in $L(E,F)$. If $Y \in m^2(E,F)$, it is called a 2-lifting of ξ. If Y is a lifting of ξ, we obtain a 2-lifting by defining

$Y_m = Y$ if $\|Y\|_{(2)} \leq m$, $Y = 0$ else, and then choosing a $\delta \in {}^*\mathbb{N} \setminus \mathbb{N}$ such that

$$\lim_{m \to \infty} {}^{o}E\left(\int_0^1 \|Y_m(t,\omega)\|^2_{(2)} dt\right) = {}^{o}E\left(\int_0^1 \|Y_\delta(t,\omega)\|^2_{(2)} dt\right).$$

Then Y_δ is a 2-lifting of ξ, as is easily seen. Hence every $\xi \in M(L_{(2)}(H,K))$ has a 2-lifting.

We have the simple

Lemma 4: Let $X \in M[L_{(2)}(H,K)]$, and let $Y, Y' \in m^2(E,F)$ be 2-liftings of X. Then there is a set Ω' of Loeb-measure one such that

$$\int_0^t Y(s,\omega) d\chi(s,\omega) = \int_0^t Y'(s,\omega) d\chi(s,\omega)$$

for all t and all $\omega \in \Omega'$.

Proof: By Doob's inequality:

$$E\left(\sup_{t \leq 1} \left\| \int_0^t (Y-Y') d\chi(s,\omega) \right\|^2\right) \leq 4E\left(\left\| \int_0^1 (Y-Y') d\chi \right\|^2\right)$$

$$= 4E\left(\int_0^1 \|Y-Y'\|^2_{(2)} ds\right) \approx 0.$$

By Lemma 4, and the fact that if $Y \in m^2(E,F)$, $\int Y d\chi$ is near-standard (Theorem I-4), the following definition makes sense:

Definition 5: Let $\xi \in M[L_{(2)}(H,K)]$. Then the stochastic integral $\int \xi dW$ is defined by

$$\int_0^t \xi dW = {}^\circ\int_0^t Y d\chi ,$$

where Y is a 2-lifting of ξ.

From Theorem I-4 it follows that $\int \xi dW$ is S-continuous. Observe that we have defined the stochastic integral without using the apparatus of section 3.

4. The Standard Approach to Stochastic Integration.

We briefly review the standard theory for stochastic integration in Hilbert spaces (see Kuo [7].) Let (i,H,B) be an abstract Wiener-space with a Brownian motion W, and let K be a Hilbert space. By modifying B a little if necessary, one may prove that a bounded linear operator from B to K is a Hilbert-Schmidt operator when regarded as an operator from H to K, and that $L(B,K)$ is dense in $L_{(2)}(H,K)$.

A process $\xi \in M[L(B,K)]$ is called a <u>simple process</u> if there is a finite sequence $0 = t_o < t_1 < .. < t_n = 1$, such that each ξ_{t_i} is a simple function, and ξ is constant on each interval $[t_j,t_{j+1})$. The stochastic integral of this simple process with respect to W is the process

$$J_\xi(t) = \sum_{j=o}^{n-1} \xi(t_j)(W(t_{j+1}\wedge t) - W(t_j\wedge t)).$$

J_ξ is a continuous martingale with

$$E(J_\xi(t)) = 0$$

and

$$E(\|J_\xi(t)\|^2) = E(\int_0^t \|\xi(s)\|^2_{(2)} ds).$$

Given a process $\xi \in M[L_{(2)}(H,K)]$, one may show that there exists a sequence $\{\xi_n\}$ of simple processes in $M[L(B,K)]$ such that

$$\lim_{n\to\infty} E\int_0^1 \|\xi_n(s)-\xi(s)\|^2_{(2)}ds = 0,$$

and hence that $\{J_{\xi_n}\}$ converges in mean square to some process J_ξ, i.e.

$$\lim_{n\to\infty} E(\|J_{\xi_n}(t)-J_\xi(t)\|^2) = \lim_{n\to\infty}(\int_0^t \|\xi_n(s)-\xi(s)\|^2_{(2)}ds) = 0$$

As the stochastic integral $\int \xi dW$ we choose such a process J_ξ with continuous sample paths. Then $\int \xi dW$ is a martingale with

$$E(\int_0^t \xi dW) = 0 \quad\text{and}\quad E(\|\int_0^t \xi dW\|^2) = E(\int_0^t \|\xi(s)\|^2_{(2)}ds)$$

5. Equivalence of the two Approaches.

Theorem 6: Let (i,H,B) be an abstract Wiener space, and let W be standard part of an Anderson process χ. If $\xi\in M[L_2(H,K)]$, then the two definitions of $\int \xi dW$ agree.

Proof: Assume first that $\xi\in M[L(B,K)]$ is a simple process jumping in $0=t_0<t_1<t_2<\ldots<t_n=1$, and let $L\in\mathbb{N}$ be such that $\|\xi(t,\omega)\|_{B,K}<L$ for all (t,ω). We may then find a simple lifting $Y\in m^2(E,F)$ jumping in points $0=\hat{t}_0<\hat{t}_1<\ldots<\hat{t}_n=1$, where $\hat{t}_j\approx t_j$ for each $j\leq n$, and with $\|Y(t,\omega)\|_{(E,|\cdot|),F}<L$ for all (t,ω). Then the expressions

$$\sum_{j=0}^{n-1} \xi(t_j)[W(t_{j+1}\wedge t)-W(t_j\wedge t)]$$

and

$$^{\circ}(\sum_{j=0}^{n-1} Y(\hat{t}_j)[\chi(\hat{t}_{j+1}\wedge t)-\chi(\hat{t}_j\wedge t)])$$

are equal a.e. since Y is a 2-lifting of ξ and $|\cdot|$-continuous.

To prove the theorem for general $\xi \in M[L_2(H,K)]$, we must prove that if $\lim_{n\to\infty} (\int_0^1 \|X_n - X\|^2_{(2)} ds) = 0$, and Y and Y_n are 2-liftings of X and X_n respectively, then

$$\lim_{n\to\infty} {}^{o}E(\|\int Y_n d\chi - \int Y d\chi\|^2) = 0$$

But

$$\lim_{n\to\infty} {}^{o}E(\|\int_0^t Y_n d\chi - \int_0^t Y d\chi\|)^2 = \lim_{n\to\infty} {}^{o}E(\int_0^t \|Y_n - Y\|^2_{(2)} ds) = \lim_{n\to\infty} E(\int_0^t \|X_n - X\|^2_{(2)} ds) = 0,$$

and the theorem is proved.

Hence we have two ways of constructing the same theory. The nonstandard theory has the following two advantages:

1. The stochastic integral has a simple, intuitive definition as pathwise Stieltjes integral.

2. The Brownian motion has a natural construction as a random walk on a hyperfinite dimensional linear space. Thus we avoid the extension of H to B, Gross' theorem, and the rather bothersome theory of the relationship between H and B and their linear maps. By the way, Theorem 6 implies that it is the same which measurable norm we use; we will always get the same stochastic integral.

A drawback with the nonstandard theory is perhaps that it needs the machinery of liftings, but this is a central notion of nonstandard measure theory. Also the nonstandard method applies only to Brownian motions generated by Anderson processes, but in view of the "internal transformation"-theory of Keisler [6], and the representation theorems of [10], this is probably not of great importance.

6. Applications to Partial Stochastic Differential Equations.

We shall now show how the theory developed in Part II can be combined with the result of Section I-4, to yield existence results for solutions of partial stochastic differential equations. The equations we shall study will be of the form

(*) $\quad dx(t) = Ax(t)dt + \xi(t,x(t))dW(t), \; x(0) = x_o$

where A is the infinitesimal generator of a strongly continuous semigroup T_t on our separable Hilbert space K, W is a Brownian motion on H, and $\xi(t,u)$ are Hilbert-Schmidt mappings from H to K. The equation (1) may be interpreted in several ways, two of them are

(**) $\quad$ weakened solution: $x(t) = x_o + A\int_0^t x(s)ds + \int_0^t \xi(s,x(s))dW(s)$

and

(***) $\quad$ mild integral solution: $x(t) = T_t x_o + \int_0^t T_{t-s} \circ \xi(s,x(s))dW(s).$

We shall prove

<u>Theorem 7</u>: Let H and K be real, separable Hilbert-spaces, and let W be the standard part of an Anderson process on H. Let $A : K \to K$ be the infinitesimal generator of a strongly continuous semigroup T_t, and let $\xi : [0,1] \times K \to L_{(2)}(H,K)$ be measurable and assume that there is an $L \in \mathbb{N}$ such that $\|\xi(t,u) - \xi(t,v)\|_{(2)} \leq L\|u-v\|$ for all t,u,v, and that $\int_0^1 \|\xi(t,0)\|^2_{(2)} < \infty$. Finally, let x_o be an $L(G_o)$-measurable, square integrable initial condition. Then

(*) $\quad dx(t) = Ax(t)dt + \xi(t,x(t)dW(t), \; x(0) = x_o$

has a continuous weakened and mild integral solution x.

Proof: We first prove that (*) has a mild integral solution. Let χ be the Anderson process that generates W, and let E be the hyperfinite dimensional subspace of ${}^*H$ where χ lives. Let $\hat{\xi}: [0,1] \to C(K, L_{(2)}(H,K))$ be defined by

$$\hat{\xi}(t)(u) = \xi(t,u),$$

and let $\hat{\Xi}: T \to {}^*C[{}^*K, L_{(2)}(E, {}^*K)]$ be a lifting of $\hat{\xi}$ in the closed-open topology. Define $\Xi: T \times {}^*K \to L_{(2)}(E, {}^*K)$ by

$$\Xi(t,u) = \hat{\Xi}(t)(u).$$

We choose $\hat{\Xi}$ such that Ξ is uniformly Lipschitz continuous in the second variable, and $\|\Xi(t,0)\|^2_{(2)}$ is S-integrable over T. Let X_0 be a G_0-measurable 2-lifting of x_0.

By Theorem I-6 is then

$$X_t = {}^*T_t X_0 + \sum_0^t {}^*T_{t-\Delta t-s}\Xi(s, X(s))\Delta\chi(s)$$

near-standard and S-continuous.

Now ${}^*T_t X_0 = T_t x_0 L(P)$-a.e., and ${}^*T_{t-\Delta t-s}\Xi(s,X(s))$ equals a 2-lifting of $T_{t-s}\xi(s, {}^oX(s))$ on a set of measure one, and thus

$${}^oX(t) = T_t {}^oX_0 + \int_0^t T_{t-s}\xi(s, {}^oX(s))dW,$$

and $x(t) = {}^oX(t)$ is a mild integral solution of (*). Since X is S-continuous, x is continuous.

Since there is a general theorem (see Chojnowska-Michalik [3]) that a process is a weakened solution if and only if it is a mild integral solution, we only sketch the nonstandard argument for x being a weakened solution leaving the details to the reader:

We first prove that $\int_0^t x(s)ds$ is in the domain of A:

$$A\int_0^t T_s x_o ds = \lim_{r\to\infty} \frac{T_r - I}{r} \int_0^t T_s x_o ds = \lim \frac{1}{r}\left(\int_t^{t+r} T_s x_o ds - \int_0^r T_s x_o ds\right) = T_t x_o - x_o.$$

To see that $\int_0^t [\int_0^s T_{s-r}\xi(r,x(r))dW(r)]ds$ is in the domain of A, we first notice that

$$\int_0^t [\int_0^s T_{s-r}\xi(r,x(r))dW(r)]ds = {}^{\circ}\sum_{s=o}^{t} [\sum_{r=o}^{s} {}^{*}T_{s-r-\Delta t}\Xi(r,X(r))\Delta\chi(r)]\Delta t$$

$$= {}^{\circ}\sum_{r=o}^{t} [\sum_{s=r}^{t} {}^{*}T_{s-r-\Delta t}\Xi(r,X(r))\Delta t]\Delta\chi(r) = \int_0^t [\int_r^t T_{s-r}\xi(r,x(r))ds]dW(r)$$

Since by a calculation similar to the one above $A\int_r^t T_{s-r}\xi(r,x(r))ds = T_{t-r}\xi(r,x(r)) - \xi(r,x(r))$, it is enough to prove that if $\Phi(t) \in D(A)$ and $\varphi, A\varphi \in M[L_{(2)}(H,K)]$, then $\int_0^t \varphi dW \in D(A)$ and $A\int_0^t \varphi dW = \int_0^t A\varphi dW$. This is done as follows:

Since $\varphi, A\varphi : [0,1]\times\Omega \to L_{(2)}(H,K)$, we may define $\tilde{\varphi} : [0,1]\times\Omega \to L_{(2)}(H,G(A))$ (where $G(A)$ is the graph of A) by $\tilde{\varphi}(t,\omega)(u) = (\varphi(t,\omega)(u), A\varphi(t,\omega)(u))$. Let $\tilde{\Phi}$ be a 2-lifting of $\tilde{\varphi}$, $\tilde{\Phi} : T\times\Omega \to L_{(2)}({}^{*}H, {}^{*}G(A))$, and let Φ be the first component of $\tilde{\Phi}$. Then

$${}^{\circ}\int \Phi d\chi = \int \varphi dW$$

$$\text{and} \quad {}^{\circ *}A\int \Phi d\chi = {}^{\circ}\int {}^{*}A\Phi d\chi = \int A\varphi dW.$$

Hence $(\int \Phi d\chi, {}^{*}A\int \Phi d\chi) \in {}^{*}G(A)$ is near-standard, and since $G(A)$ is closed $({}^{\circ}\int \Phi d\chi, {}^{\circ *}A\int \Phi d\chi) \in G(A)$, and hence $\int \varphi dW \in D(A)$ and $A\int \varphi dW = \int A\varphi dW$.

This proves that $\int_0^t x(s)ds \in D(A)$, and we also get

$$A\int_0^t x(s)ds = A(T_t x_o) + A\int_0^t [\int T_{s-r}\xi(r,x(r))ds]dW(r)$$

$$= T_t x_o - x_o + \int_0^t [T_{t-r}\xi(r,x(r)) - \xi(r,x(r))]dW(r).$$

Hence

$$x_o + A\int_0^t x(s)ds + \int_0^t \xi(s,x(s))dW(s) = T_t x_o + \int_0^t T_{t-r}\xi(r,x(r))dW(r) = x(t),$$

which proves that x also is a weakened solution, and hence the theorem.

Theorems of this sort have been obtained by standard methods; see Chojnowska-Michalik [3] and references therein. Perhaps more interesting are the results obtained by Bensoussan and Teman [2], Pardoux [13], and Chow [4] using the Lions approach, but as yet no satisfactory nonstandard theory for their kind of equations has been developed.

References

0. R.M. Anderson: A Nonstandard Representation of Brownian Motion and Itô Integration. Israel J. Math. 25 (1976) pp. 15-46.

1. R.M. Anderson: Star-finite Probability Theory. Ph.D.-thesis, Yale University, 1977.

2. A. Bensoussan and R. Teman: Equation aux Derivées Partielles Stochastiques, Israel J. Math. 11 (1972)

3. A. Chojnowska-Michalik: Stochastic Differential Equations in Hilbert Spaces, in Z. Ciesielski (ed): Probability Theory. Banach Center Publications, Vol. 5, PWN, 1979.

4. P.-L. Chow: Stochastic Partial Differential Equations in Turbulence Related Problems, in A.T. Bharucha-Reid (ed): Probabilistic Analysis and Related Topics 1, Academic Press 1978,

5. L. Gross: Abstract Wiener Spaces, Proc. 5th. Berkeley Sym. Math. Stat. Prob. 2 (1965), pp 31-42.

6. H.J. Keisler: An Infinitesimal Approach to Stochastic Analysis, Preliminary Version, 1978. (To app. Mem. AMS)

7. H.-H. Kuo: Gaussian Measures in Banach Spaces, LNM 463, Springer-Verlag, 1975.

8. T.L. Lindstrøm: Hyperfinite Stochastic Integration I: The Nonstandard Theory. Math.Scand.46, pp265-292.

9. T.L. Lindstrøm: Hyperfinite Stochastic Integration II: Comparison with the Standard Theory. Math.Scand.46,pp293-314.

10. T.L. Lindstrøm: Hyperfinite Stochastic Integration III: Hyperfinite Representations of Standard Martingales. Math. Scand. 46 (1980), pp 315-332.

11. T.L. Lindstrøm: A Loeb-measure Approach to Theorems by Prohorov, Sazonov, and Gross. Trans AMS 269 (1982).

12. P.A. Loeb: An Introduction to Nonstandard Analysis and Hyperfinite Probability Theory, in A.T. Bharucha-Reid (ed): Probabilistic Analysis and Related Topics 2, Academic Press, 1979, pp 105-142.

13. E. Pardoux: Equations aux derivées partielles stochastiques monotones, C.R. Acad. Sci. Paris, Ser. A, 275(1972).

14. M. Reed and B. Simon: Methods of Modern Mathematical Physics II, Academic Press, 1975.

15. K.D. Stroyan and W.A.J. Luxemburg: Introduction to the Theory of Infinitesimals, Academic Press, 1976.

16. D.N. Hoover and E.A. Perkins: Nonstandard Construction of the Stochastic Integral and Applications to Stochastic Differential Equations, I-II, To appear Trans. AMS.

17. T.L. Lindstrøm: The Structure of Hyperfinite Stochastic Integrals. To appear Zeit. f. Wahr. Theorie.

+) Current address: Department of Mathematics, University of Wisconsin, Madison, Wisconsin 53706.

Stochastic Processes and Nonstandard Analysis

Edwin Perkins
Department of Mathematics
University of British Columbia
Vancouver, B.C. V6T 1Y4
Canada

1. Introduction

This article presents some recent applications of nonstandard analysis, and in particular the "Loeb space", in probability theory. It is by no means a complete survey of results, but rather concentrates on the topics of stochastic integration, stochastic differential equations and local time, thus reflecting the author's own interests. We assume a basic knowledge of nonstandard analysis and the fundamental construction of Loeb [21]. For example, the details of ω_1-saturation arguments are usually omitted. All the necessary prerequisites may be found in the introductory article of Loeb [22] (especially sections IV and V). In fact the present work may be regarded as a natural sequel to [22]. Only a minimal amount of probability theory is assumed in all but the last section, where some knowledge of the general theory of processes would be helpful (Meyer [24] is a good source). Generality is often sacrificed for the sake of presenting what we hope is a clear and coherent story. Proofs are often only sketched or omitted entirely, but references to complete proofs will always be given.

Section 2 presents some preliminary material including Anderson's [1] construction of Brownian motion and a description of the particular Loeb space that will be used in all but the last section. In section 3 we present the nonstandard construction of the Itô integral originally due to Anderson [1]. Proofs are presented in full to illustrate the basic techniques and also because the class of integrands considered here is slightly larger than that studied by Anderson. The simplest results of Keisler [14] on "strong" existence theorems for Itô equations are presented in section 4, and section 5 describes some of the author's results on Brownian local time. Finally, in section 6 the results of sections 3 and 4 are extended to stochastic integration with respect to semimartingales (see Hoover and Perkins [11], [12] and Lindstrøm [17], [18], [19]).

Hopefully sections 4 and 6 will provide good introductions to the material in Keisler [14] and Hoover and Perkins [11], [12], respectively. The references to Keisler's fundamental paper refer to the latest available preprint and as a result our numbering of the theorems may not coincide exactly with the final version. Much of the material presented here will be studied in greater detail in the forthcoming book of Stroyan and Bayod [32]. Those interested in further reading may find other articles of interest in Keisler's exhaustive bibliography [14].

As usual we work in an ω_1-saturated enlargement of a superstructure containing the real numbers. We close with a few results and notations not given in [22]. $(\Omega, L(A), L(\nu))$ denotes the Loeb space constructed from an internal measure space, (Ω, A, ν). That is, ${}^\circ\nu$ is extended to a unique σ-additive measure on $\sigma(A)$, the standard σ-algebra generated by A, and $(\Omega, L(A), L(\nu))$ denotes the completion of the resulting measure space. We also need the following simple Fubini theorem of Keisler [14, Th. 1.14b].

Theorem 1. If (Ω_i, A_i, ν_i) $i = 1,2$ are internal finite (${}^\circ\nu_i(\Omega_i) < \infty$) measure spaces and $f:\Omega_1 \times \Omega_2 \to \mathbb{R}$ is bounded and $L(A_1 \times A_2)$ - measurable, then

(i) $\omega_2 \to f(\omega_1,\omega_2)$ is $L(A_2)$ - measurable for $L(\nu_1)$ - a.a. ω_1,

(ii) $\omega_1 \to \int f(\omega_1,\omega_2)dL(\nu_2)$ is $L(A_1)$ - measurable,

(iii) $\int\int f(\omega_1,\omega_2)dL(\nu_1 \times \nu_2) = \int(\int f(\omega_1,\omega_2)dL(\nu_2))dL(\nu_1)$. □

(Anderson [1] showed that $L(A_1) \times L(A_2) \subset L(A_1 \times A_2)$ and an example of Hoover shows that one does not have equality in general.)

Finally, we use the notation $I(A)$ to denote the indicator function of the set A.

2. Preliminaries

Choose an infinite η in ${}^*\mathbb{N}$ and let $T = \{i\eta^{-1} | i=0,\dots,\eta^2\}$. We often write Δt for η^{-1}. Our sample space is the internal product space $\Omega = \{-1,1\}^T$, which we equip with the algebra, A, of internal subsets and the internal probability measure, $\overline{P}$, assigning equal mass $(2^{-\eta^2-1})$ to each point of Ω. The Loeb space constructed from $(\Omega, A, \overline{P})$ is denoted by (Ω, F, P). Elements of T are denoted by $\underline{s}, \underline{t}, \dots$ For each $\underline{t}$ in T, $A_{\underline{t}}$ is the internal σ-algebra generated by the coordinate mappings $\omega \to \omega_{\underline{s}}$ for $\underline{s} \le \underline{t}$. A non-decreasing family of sub-σ-algebras of F, $\{F_t | t \in [0,\infty)\}$ is defined by

$$F_t = ({}_\circ\cap_{\underline{s}>t}\sigma(A_{\underline{s}})) \vee N,$$

where N denotes the class of P-null sets in F. It is easy to see that $F_t = \cap_{s>t} F_s$ and hence $\{F_t\}$ satisfies the "usual hypotheses" of Meyer et al. We use $\overline{E}$ to denote internal expectation with respect to $\overline{P}$, while E denotes expectation with respect to P.

Note that if $\omega = (\omega_{\underline{s}})_{\underline{s}\in T} \in \Omega$, then $\{\omega_{\underline{s}} | \underline{s} \in T\}$ are independent random variables on (Ω, F, P), each taking on the values ± 1 with equal probability. Anderson's [1] infinitesimal random walk is defined by

$$X(\underline{t},\omega) = \sum_{0<\underline{s}\leq\underline{t}} \omega_{\underline{s}}(\Delta t)^{\frac{1}{2}} .$$

We wish to use the $*\mathbb{R}$-valued, internal process, X, to construct a real-valued stochastic process on the Loeb space (Ω,F,P). To this end we now consider the space, $C(L,\mathbb{R}^n)$, of $\mathbb{R}^n$-valued continuous functions on L $(L \subset \mathbb{R}^k)$ from a non-standard viewpoint. We give $C(L,\mathbb{R}^n)$ the compact-open topology (i.e., the topology of uniform convergence on compacts).

Definition. Let Λ be an internal subset of $*\mathbb{R}^k$. An internal function $F:\Lambda\to *\mathbb{R}^n$ is S-continuous (on Λ) if and only if whenever $\lambda_1 \approx \lambda_2$ are in $ns(\Lambda)$, $°F(\lambda_1) = °F(\lambda_2) \in \mathbb{R}^n$.

It is easy to see that if F is S-continuous on Λ $(\Lambda \subset *\mathbb{R}^k)$, there is a unique continuous function $f:st(\Lambda) = \{°\lambda \mid \lambda \in ns(\Lambda)\} \to \mathbb{R}^n$ satisfying $f(°\lambda) = °F(\lambda)$ for each $\lambda \in ns(\Lambda)$. We write $f = st(F)$. An internal process, $X:\Lambda \times \Omega \to *\mathbb{R}^n$ is S-continuous, if and only if $X(\cdot,\omega)$ is S-continuous a.s., and in this case we may define an $\mathbb{R}^n$-valued stochastic process with continuous paths on $st(\Lambda)$, $st(X)$, by

$$st(X)(\lambda,\omega) = \begin{cases} st(X(\cdot,\omega))(\lambda) & \text{if } X(\cdot,\omega) \text{ is S-continuous} \\ 0 & \text{otherwise.} \end{cases} \quad \square$$

(In the above definition $ns(\Lambda)$ denotes the near-standard points in Λ and $°$ denotes the standard part map, as usual.)

Proposition 2 (Keisler [14, Prop. 1.17] or Stroyan and Luxemburg [33, Th. 8.44]). If $L \subset \mathbb{R}^k$, then $F \in *C(L,\mathbb{R}^n)$ is S-continuous if and only if it is near-standard in $*C(L,\mathbb{R}^n)$ and in this case $st(F)$ is the standard part of F in $C(L,\mathbb{R}^n)$. $\square$

We are now ready to present Anderson's [1] construction of Brownian motion. A proof may also be found in Loeb [22, V.C, Th. 3,4].

Theorem 3. The infinitesimal random walk, X, is S-continuous. The process, $B = st(X)$, is a continuous Brownian motion on (Ω,F,P,F_t). That is, for all $s < t$, $B(t)$ is F_t-measurable, $B(t) - B(s)$ is independent of F_s and has a normal distribution with mean zero and variance $t - s$, and $t \to B(t,\omega)$ is continuous for each ω. $\square$

3. Itô Integration

To motivate the nonstandard approach to Itô integration we first recall a nonstandard construction of the Lebesgue integral due to Anderson [1].

Let C denote the algebra of internal subsets of T and let λ be the internal measure on (T,C) assigning mass Δt to each singleton. We abuse notation slightly by writing $(ns(T),L(C),L(\lambda))$ for the "restriction" of the Loeb space

$(T, L(C), L(\lambda))$ to the measurable subset, $ns(T)$. We note that if $\mathcal{B}$ denotes the Lebesgue-measurable subsets of $[0,\infty)$ and m is Lebesgue measure then

(0) $\quad {}^\circ : (ns(T), L(C), L(\lambda)) \to ([0,\infty), \mathcal{B}, m)$ is measurable and measure-preserving.

(First check intervals in $[0,\infty)$ and then apply the usual monotone class arguments. Another argument is given at the end of section IV in Loeb [22].) Therefore if $f:[0,\infty) \to \mathbb{R}$ is bounded and Lebesgue measurable, then $\underline{t} \to f({}^\circ\underline{t})$ is $L(C)$-measurable on $ns(T)$ and hence, by Loeb's lifting theorem (Loeb [22, IV.A, Th. 9]), there is an internal, (standardly) bounded function, $F: T \to {}^*\mathbb{R}$, such that ${}^\circ F(\underline{t}) = f({}^\circ\underline{t})$ $L(\lambda)$-a.e. on $ns(T)$. (Although the lifting theorem in [22] holds only on finite measure spaces, one may use it to obtain liftings F_n such that ${}^\circ F_n(\underline{t}) = f({}^\circ\underline{t})$ $L(\lambda)$-a.e. on $T \cap {}^*[0,n]$ and then use saturation.) Therefore for each $\underline{t} \in ns(T)$ we have

$$\begin{aligned} {}^\circ \sum_{\underline{s}<\underline{t}} F(\underline{s})\Delta t &= {}^\circ\!\int I(\underline{s} < \underline{t}) F(\underline{s}) d\lambda \\ &= \int I(\underline{s} < \underline{t})\, {}^\circ F(\underline{s}) dL(\lambda) \quad \text{(Loeb [22, IV.A, Th. 4])} \\ &= \int I({}^\circ\underline{s} \le {}^\circ\underline{t}) f({}^\circ\underline{s}) dL(\lambda) \end{aligned}$$

$$(1) \quad \therefore\ {}^\circ \sum_{\underline{s}<\underline{t}} F(\underline{s})\Delta t = \int_0^{{}^\circ\underline{t}} f(s) ds \qquad \text{(by (0))}.$$

Hence, to construct $\int_0^t f ds$ we first find an appropriate lifting, F, of f and then take the standard part of an internal Riemann sum. The same procedure may be used to construct the Itô integral.

The Itô integral, $\int_0^t h(s,\omega) dB(s,\omega)$, is defined (standardly) whenever

$$h \in I = \{h:[0,\infty) \times \Omega \to \mathbb{R} \mid h \text{ is measurable, } h(s,\cdot) \text{ is } F_s\text{-measurable for each } s \ge 0 \text{ and } \int_0^t h^2(s,\omega) ds < \infty \text{ for all } t > 0 \text{ a.s.}\}$$

(see McKean [23]). The nonstandard approach of Anderson [1] gives another construction of this integral, as we now show.

Notation. 1) $\overline{I} = \{H: T \times \Omega \to {}^*\mathbb{R} \mid H$ is internal, $H(\underline{t},\cdot)$ is $A_{\underline{t}}$-measurable for all $\underline{t} \in T$, $H(\cdot,\omega)^2 I(\cdot < m)$ is S-integrable on $(T, \overline{C}, \lambda)$ for each $m \in \mathbb{N}$ and for a.a.ω, and $\sup_{(\underline{t},\omega)} |H(\underline{t},\omega)| \in {}^*\mathbb{R}\}$.

2) If $H: T \times \Omega \to {}^*\mathbb{R}$ is internal, let

$$\int_0^{\underline{t}} H(\underline{s},\omega) dX(\underline{s},\omega) = \sum_{\underline{s}<\underline{t}} H(\underline{s},\omega) \Delta X(\underline{s},\omega),$$

where $\Delta X(\underline{s},\omega) = X(\underline{s} + \Delta t, \omega) - X(\underline{s},\omega)$, and let

$$\int_0^{\underline{t}} H(\underline{s},\omega) d\lambda = \int I(\underline{s} < \underline{t}) H(\underline{s},\omega) d\lambda = \sum_{\underline{s}<\underline{t}} H(\underline{s},\omega)\Delta t.$$

Definition. An internal process $H: T \times \Omega \to {}^*\mathbb{R}$ (i.e., H is an internal function) is a lifting of a stochastic process $h:[0,\infty) \times \Omega \to \mathbb{R}$ if ${}^\circ H(\underline{t},\omega) = h({}^\circ\underline{t},\omega)$ $L(\lambda \times \overline{P})$-a.s. on $ns(T) \times \Omega$. We say that an internal process, H, is $A_{\underline{t}}$-adapted if $H(\underline{t},\cdot)$ is $A_{\underline{t}}$-measurable for each $\underline{t} \in T$. □

The following result is our first example of a lifting theorem, that is, a result which associates to a given standard process, an internal process with analogous properties. A quick glance through Keisler's work will convince one of the importance of this notion in nonstandard stochastic analysis. It is for this reason that we give a complete proof.

Proposition 4. Every process in I has a lifting in $\overline{I}$.

Proof. Assume $h \in I$. There is a sequence of functions, $\{h_n\}$, in I of the form

$$(2)\qquad h_n(t,\omega) = \sum_{i=0}^{N_n} h_i^n(\omega) I_{[t_i^n, t_{i+1}^n)}(t), \qquad 0 = t_0^n < t_1^n < \ldots < t_{N_n+1}^n \le \infty,$$

such that $|h_n| \le C_n$ $(C_n \in \mathbb{R})$,

$$(3)\qquad P((\int_0^n (h - h_n)^2 (s,\omega) ds) \ge 2^{-(n+1)}) < 2^{-(n+1)},$$

and therefore,

$$(4)\qquad P((\int_0^n (h_m - h_n)^2 (s,\omega) ds) \ge 2^{-n}) < 2^{-n}, \quad \forall m \ge n.$$

(The existence of such a sequence follows easily from Doob [9, pp. 439-442].) Fix n, $i \le N_n$ and choose a sequence, $\{\underline{s}_k | k \in \mathbb{N}\}$, in T such that $°\underline{s}_k$ decreases to t_i^n. As h_i^n is a.s. equal to a $\sigma(A_{\underline{s}_k})$-measurable random variable, we may apply Theorem 9 of IV.A of Loeb [22] to obtain an internal $A_{\underline{s}_k}$-measurable random variable H_k such that $°H_k = h_i^n$ a.s. and $|H_k| \le C_n$. A routine saturation argument now produces an internal random variable H_i^n such that $°H_i^n = h_i^n$ a.s., $|H_i^n| \le C_n$, and H_i^n is $A_{\underline{t}_i^n}$-measurable for some $\underline{t}_i^n \approx t_i^n$. Let

$$H_n(\underline{t},\omega) = \sum_{i=0}^{N_n} H_i^n(\omega) I_{[\underline{t}_i^n, \underline{t}_{i+1}^n)}(\underline{t}),$$

where $°\underline{t}_{N_n+1}^n = t_{N_n+1}^n$. Then (4) and the previous construction of the Lebesgue integral (see (1)) show that

$$\overline{P}\,((\int_0^n (H_m - H_n)^2 (\underline{s},\omega) d\lambda)^{\frac{1}{2}} > 2^{-n}) < 2^{-n}, \quad \forall m \ge n.$$

By ω_1-saturation we may extend $\{H_n | n \in \mathbb{N}\}$ to an internal sequence indexed by $*\mathbb{N}$ and hence obtain an internal process, H_γ, (for some $\gamma \in *\mathbb{N} - \mathbb{N}$) that is $A_{\underline{t}}$-adapted bounded in absolute value by C_γ and satisfies

$$(5)\qquad \overline{P}\,((\int_0^n (H_\gamma - H_n)^2 (\underline{s},\omega) d\lambda)^{\frac{1}{2}} > 2^{-n}) < 2^{-n}, \quad \forall n \le \gamma.$$

(3) and (5), together with the fact that H_n is a lifting of h_n, show that $H_\gamma \in \overline{I}$ and H_γ is a lifting of h. □

Having found a lifting, H, of $h \in I$, we wish to construct $\int_0^t h(s) dB(s)$ by taking the standard part of $\int_0^{\underline{t}} H(\underline{s}) dX(\underline{s})$. To do this we show that $\int_0^{\underline{t}} H(\underline{s}) dX(\underline{s})$ is S-continuous for $H \in \overline{I}$ and that the standard part of this internal process does not depend on the choice of lifting. The latter problem is the simpler.

Definition. An internal, ${}^*\mathbb{R}$-valued, $A_{\underline{t}}$-adapted process, $M(\underline{t},\omega)$, is an $(A_{\underline{t}})$-martingale if $\overline{E}(|M(\underline{t})|) \in {}^*\mathbb{R}$ for $\underline{t} \in T$ and $\overline{E}(M(\underline{t}+\Delta t)|A_{\underline{t}}) = M(\underline{t})$ $\overline{P}$-a.s. An internal mapping, $V:\Omega \to T \cup \{\infty\}$, is an $(A_{\underline{t}})$-stopping time if $\{V = \underline{t}\} \in A_{\underline{t}}$ for each $\underline{t}$ in T. □

Note that if $H \in \overline{I}$ then

$$\overline{E}(H(\underline{t},\omega)\Delta X(\underline{t},\omega)|A_{\underline{t}}) = 0 \qquad \overline{P}\text{-a.s.}$$

(recall that $\sup_{(\underline{t},\omega)} |H(\underline{t},\omega)| \in {}^*\mathbb{R}$). This shows that $\int_0^{\underline{t}} H(\underline{s},\omega)dX(\underline{s},\omega)$ is an $(A_{\underline{t}})$ martingale.

Notation. $\xrightarrow{P}$ denotes convergence in probability.

Proposition 5. (a) If H_1, $H_2 \in \overline{I}$ satisfy ${}^\circ H_1(\underline{t},\omega) = {}^\circ H_2(\underline{t},\omega)$ $L(\lambda \times \overline{P})$-a.s. on $ns(T) \times \Omega$, then

$$\int_0^{\underline{t}} H_1(\underline{s},\omega)dX(\underline{s},\omega) \approx \int_0^{\underline{t}} H_2(\underline{s},\omega)dX(\underline{s},\omega) \qquad \text{for all } \underline{t} \in ns(T) \text{ a.s.}$$

(b) If H_n, $H_\infty \in \overline{I}$ $(n \in \mathbb{N})$ satisfy

$${}^\circ\int_0^m (H_n(\underline{s},\omega) - H_\infty(\underline{s},\omega))^2 d\lambda \xrightarrow{P} 0 \quad \text{as } n \to \infty, \quad \forall\, m \in \mathbb{N}$$

then

$${}^\circ\max_{\underline{t}\le m} \left|\int_0^{\underline{t}} H_n(\underline{s},\omega)dX(\underline{s},\omega) - \int_0^{\underline{t}} H_\infty(\underline{s},\omega)dX(\underline{s},\omega)\right| \xrightarrow{P} 0 \quad \text{as } n \to \infty, \quad \forall\, m \in \mathbb{N}.$$

Proof. Let $H \in \overline{I}$, and for $b > 0$ define

$$T(b) = \min\{\underline{t} \mid \int_0^{\underline{t}+\Delta t} H^2(\underline{s},\omega)d\lambda > b\} \qquad (\min \phi = \infty).$$

Since $\int_0^{\underline{t}} H^2(\underline{s},\omega)d\lambda$ is $A_{\underline{t}-\Delta t}$-measurable for $\underline{t} \ge \Delta t$, $T(b)$ is an $(A_{\underline{t}})$-stopping time. Note that for $\varepsilon > 0$ and $m \in \mathbb{N}$,

$$\begin{aligned}
\overline{P}(\max_{\underline{t}\le m} |\int_0^{\underline{t}} H(\underline{s},\omega)dX(\underline{s},\omega)| > \varepsilon) &\le \overline{P}(\max_{\underline{t}\le m} |\int_0^{\underline{t}} H(\underline{s},\omega)dX(\underline{s},\omega)| > \varepsilon,\ T(\varepsilon^3) \ge m) \\
&\qquad + \overline{P}(T(\varepsilon^3) < m) \\
&\le \overline{P}(\max_{\underline{t}\le m\wedge T(\varepsilon^3)} |\int_0^{\underline{t}} H(\underline{s},\omega)dX(\underline{s},\omega)| > \varepsilon) + \overline{P}(T(\varepsilon^3) < m) \\
&\le \varepsilon^{-2}\overline{E}((\int_0^{T(\varepsilon^3)} H(\underline{s},\omega)dX(\underline{s},\omega))^2) + \overline{P}(T(\varepsilon^3) < m)
\end{aligned}$$

(maximal martingale inequality, see Doob [9, p. 314])

$$\begin{aligned}
&= \varepsilon^{-2}\overline{E}(\sum_{\underline{s}<T(\varepsilon^3)} H(\underline{s},\omega)^2\Delta t) + \varepsilon^{-2}\overline{E}(2\sum_{\underline{s}<\underline{t}<T(\varepsilon^3)} H(\underline{s},\omega)H(\underline{t},\omega) \\
&\qquad \times \Delta X(\underline{s},\omega)\Delta X(\underline{t},\omega)) + \overline{P}(T(\varepsilon^3) < m) \\
&= \varepsilon^{-2}\overline{E}(\int_0^{T(\varepsilon^3)} H(\underline{s},\omega)^2 d\lambda) + \overline{P}(T(\varepsilon^3) < m)
\end{aligned}$$

(the last follows by the independence of $\Delta X(\underline{t},\omega)$ and $A_{\underline{t}}$).

(6) $\therefore\ \overline{P}(\max_{\underline{t}\le m} |\int_0^{\underline{t}} H(\underline{s},\omega)dX(\underline{s},\omega)| > \varepsilon) \le \varepsilon + \overline{P}(\int_0^m H(\underline{s},\omega)^2 d\lambda > \varepsilon^3)$.

For (a), set $H = H_1 - H_2$ in (6) to see that for each $\varepsilon\in(0,\infty)$ and $m \in \mathbb{N}$,

$$P(\max_{\underline{t}\le m} |\int_0^{\underline{t}} H_1(\underline{s},\omega)dX(\underline{s},\omega) - \int_0^{\underline{t}} H_2(\underline{s},\omega)dX(\underline{s},\omega)| > \varepsilon)$$

$$\le \varepsilon + P({}^\circ(\int_0^m (H_1 - H_2)^2(\underline{s},\omega)d\lambda) \ge \varepsilon^3)$$

$$= \varepsilon + P(\int_0^m {}^\circ(H_1 - H_2)^2(\underline{s},\omega)dL(\lambda) \ge \varepsilon^3) = \varepsilon$$

(We have used Keisler's Fubini Theorem (Theorem 1) in the last line.)

(b) is an immediate consequence of (6) (set $H = H_n - H_\infty$) . □

Definition. A ${}^*\mathbb{R}$-valued, $(A_{\underline{t}})$-adapted internal process, Y , is locally S-integrable if there is a sequence of $(A_{\underline{t}})$-stopping times $\{V_n\}$ such that ${}^\circ V_n$ increases to ∞ a.s. and $Y(V_n \wedge \underline{t})$ is S-integrable for all $\underline{t} \in T$ and $n \in \mathbb{N}$.

The quadratic variation of an internal process, Y , is the process

$$[Y](\underline{t}) = Y(0)^2 + \sum_{\underline{s}<\underline{t}} (\Delta Y(\underline{s}))^2 \quad (\Delta Y(\underline{s}) = Y(\underline{s}+\Delta t) - Y(\underline{s})) .$$

More generally if Y_1 and Y_2 are internal processes, let

$$[Y_1,Y_2](\underline{t}) = Y_1(0)Y_2(0) + \sum_{\underline{s}<\underline{t}} \Delta Y_1(\underline{s})\Delta Y_2(\underline{s}) . \quad □$$

Note that

$$[\int HdX](\underline{t}) = \int_0^{\underline{t}} H(\underline{s},\omega)^2 d\lambda ,$$

$$[\int H_1 dX, \int H_2 dX](\underline{t}) = \int_0^{\underline{t}} (H_1(\underline{s},\omega)H_2(\underline{s},\omega))d\lambda .$$

Theorem 6 (Hoover-Perkins [12, Th. 8.5(a)] and Lindstrøm [17, Th. 14], [18, Th. 2]).

An $(A_{\underline{t}})$-martingale, M , is S-continuous and locally S-integrable if and only if $[M]^{1/2}$ is S-continuous and locally S-integrable. □

We omit the proof of this important result which is the key to the nonstandard construction of solutions to Itô equations in the next section. The proof uses the martingale inequalitites of Burkholder, Gundy and Davis [5] as a means of comparing M and $[M]$. Other forms of Theorem 6, including a generalization of Keisler's original continuity theorem ([14, Th. 3.2]) may be found in Hoover-Perkins [12, Th. 8.5].

Corollary 7. If $H \in \overline{I}$, then $\int_0^{\underline{t}} H(\underline{s},\omega)dX(\underline{s},\omega)$ is S-continuous.

Proof. Assume first that $\sup_{(\underline{t},\omega)} |H(\underline{t},\omega)| \le n$ for some $n \in \mathbb{N}$.

Then

$$[\int HdX](\underline{t}) = \int_0^{\underline{t}} H^2(\underline{s},\omega)\lambda$$

is clearly S-continuous and locally S-integrable (take $V_n = n$). The result follows from Theorem 6.

For a general $H \in \bar{I}$, let $H_n = H\, I(|H| \le n)$.

Then for each $m \in \mathbb{N}$

$$\lim_{n\to\infty} {}^\circ\left(\int_0^m (H - H_n)^2(\underline{s},\omega)\,d\lambda\right) = 0$$

for all ω for which $H(\cdot,\omega)I(\cdot < m)$ is S-integrable with respect to λ. Proposition 5(b) implies that

$${}^\circ\max_{\underline{t}\le m} \left|\int_0^{\underline{t}} H\,dX - \int_0^{\underline{t}} H_n\,dX\right| \xrightarrow{P} 0 \qquad \text{as } n\to\infty,\ \forall m \in \mathbb{N}.$$

Therefore one gets a.s. convergence along a subsequence, and the S-continuity of $\int_0^{\underline{t}} H\,dX$ follows from the S-continuity of $\int_0^{\underline{t}} H_n\,dX$ established above. □

__Definition.__ If $h \in I$, choose a lifting, H, of h in $\bar{I}$ and define the Itô integral of h with respect to B by

$$\int_0^t h(s,\omega)\,dB(s,\omega) = \mathrm{st}\left(\int H\,dX\right)(t). \quad \square$$

Corollary 7 and Proposition 5(a) show that $\int h\,dB$ is a well-defined process with continuous paths and is independent of the choice of the lifting, H, up to null sets. It is easy to check that $\int_0^t h\,dB$ is F_t-measurable for each $t \ge 0$, i.e., $\int_0^t h\,dB$ is (F_t)-adapted. Note that Corollary 7 goes beyond establishing the continuity of the Itô integral since it applies to internal processes in $\bar{I}$ which need not be the lifting of any standard process in I. (The processes $\mathrm{st}(\int H\,dX)$, where H need not be a lifting of a standard h in I, are studied by Lindstrøm [20].)

We now show that the above definition agrees with the classical definition of the Itô integral. If $h \in I$, choose a sequence of bounded simple functions $\{h_n\}$ as in the proof of Proposition 4 so that (2) and (3) hold. Let H_n be the lifting of h_n obtained in the proof of Proposition 4. By definition we have

$$\int_0^t h_n(s,\omega)\,dB(s,\omega) = \sum_{i=0}^{N_n} h_i^n(\omega)\left(B(t\wedge t_{i+1}^n) - B(t\wedge t_i^n)\right),$$

so that our definition agrees with the classical definition of the Itô integral for bounded step functions. If $H \in \bar{I}$ is a lifting of h, then for each $m \in \mathbb{N}$ and a.a. ω,

$$\begin{aligned}
{}^\circ\int_0^m (H - H_n)(\underline{s},\omega)^2\,d\lambda &= \int_0^m {}^\circ(H - H_n)(\underline{s},\omega)^2\,dL(\lambda) \quad \text{(since } H(\cdot,\omega) \text{ is a.s. S-integrable)}\\
&= \int_0^m (h - h_n)({}^\circ\underline{s},\omega)^2\,dL(\lambda)\\
&= \int_0^m (h - h_n)(s,\omega)^2\,ds \qquad \text{(by (0))}\\
&\xrightarrow{P} 0 \qquad \text{as } n\to\infty.
\end{aligned}$$

Proposition 5(b) and the definition of the Itô integral imply

$$\sup_{t\le m} \left|\int_0^t h_n\,dB - \int_0^t h\,dB\right| \xrightarrow{P} 0 \qquad \text{as } n\to\infty \quad \forall m \in \mathbb{N}.$$

Since the same convergence result holds for the classical definition of the Itô integral (see McKean [23,2.3(7)]), the two definitions must coincide.

Note that we have only constructed the Itô integral, $\int h dB$, for the particular Brownian motion $B = st(X)$. Can such a construction also work for another Brownian motion, B', on (Ω, F, P, F_t) ? One possible solution to this problem (Hoover and Perkins [11], Lindstrøm [19]) is to first lift B' and then lift the integrand. A brief description of this approach in a more general setting will be given in section 6. Another approach to the problem (Keisler [14]) is to use the following striking property of our "adapted" Loeb space, (Ω, F, P, F_t). (A more general result is proved in [14].)

Theorem 8 (Keisler [14, Th. 4.5, 9.4]). If B' is a Brownian motion on (Ω, F, P, F_t) there is an internal, measure-preserving bijection $\phi: \Omega \to \Omega$ such that $A \in F_t$ if and only if $\phi(A) \in F_t$ (we call such an internal bijection an internal transformation on Ω) and $B'(t,\omega) = B(t,\phi(\omega))$ for all $t \geq 0$ a.s. □

An internal transformation preserves virtually all important properties of a process. We may now define the Itô integral with respect to B' by

(7) $$\int_0^t h(s,\omega) dB'(s,\omega) = \int_0^t h(s,\phi^{-1}(\cdot)) dB(s,\cdot) \circ \phi(\omega).$$

It is easy to check that this agrees with the usual definition.

One advantage of the nonstandard construction of the Itô integral is that the intuitive relationship $(dB)^2 = dt$ is replaced by the obvious equality $(\Delta X(\underline{t}))^2 = \Delta t$. This observation was used by Anderson [1] in his proof of Itô's Lemma. Other applications of the construction described in this section are presented in the next two sections.

4. Itô Equations

We describe the simplest existence theorem for Itô equations considered by Keisler in [14]. The setting is that introduced in the previous section.

Theorem 9 (Keisler [14, Th. 5.2]). Let σ, $f: [0,\infty) \times \mathbb{R} \to \mathbb{R}$ be bounded measurable functions such that $x \to \sigma(t,x)$ and $x \to f(t,x)$ are continuous for each $t \geq 0$. Then for every Brownian motion, B', on (Ω, F, P, F_t) and $y_0 \in \mathbb{R}$ there is an F_t-adapted process, $y'(t,\omega)$, with continuous paths such that

(8) $$y'(t,\omega) = y_0 + \int_0^t \sigma(s, y'(s,\omega)) dB'(s,\omega) + \int_0^t f(s, y'(s,\omega)) ds \quad \forall t \geq 0 \quad \text{a.s.}$$

Proof. We first assume B' is the Brownian motion $B = st(X)$ considered in the previous section. Our plan is to "lift" (8) to an internal difference equation which is trivial to solve and then take the standard part of this internal solution to obtain a solution of (8). Define $\hat{\sigma}: [0,\infty) \to C(\mathbb{R}, \mathbb{R})$ by $\hat{\sigma}(t)(x) = \sigma(t,x)$. Then $\underline{t} \to \hat{\sigma}({}^\circ\underline{t})$ is a measurable map on $(nS(T), L(C))$, taking values in the separable metric space $C(\mathbb{R},\mathbb{R})$. Therefore the usual lifting theorem (Th. 9 in IV.A of [22]) gives us a lifting $\hat{G}:T\to {}^*C(\mathbb{R},\mathbb{R})$ such that $st(\hat{G}(\underline{t})) = \hat{\sigma}({}^\circ\underline{t})$ $L(\lambda)$-a.s. on $ns(T)$. (We have used Proposition 2 here. Note also that, although Th. 9 of [22] applies only to finite Loeb spaces, $\hat{G}$ may be obtained by lifting $\hat{\sigma}({}^\circ\underline{t})$ on each $T\cap{}^*[0,n]$ and using saturation.) If $G(\underline{t},x) = \hat{G}(\underline{t})(x)$, then

(9) $\quad {}^\circ G(\underline{t},x) = \sigma({}^\circ\underline{t},{}^\circ x)$ for all x in $ns({}^*\mathbb{R})$, $L(\lambda)$-a.s. on $ns(T)$.

Similarly one obtains an internal process $F:T\times{}^*\mathbb{R}\to{}^*\mathbb{R}$ so that

(10) $\quad {}^\circ F(\underline{t},x) = f({}^\circ\underline{t},{}^\circ x)$ for all x in $ns({}^*\mathbb{R})$, $L(\lambda)$-a.s. on $ns(T)$.

Clearly we may assume ${}^\circ\sup_{(\underline{t},x)}(|F(\underline{t},x)| \vee |G(\underline{t},x)|) < \infty$.

Now define an internal process, $Y:T\times\Omega\to{}^*\mathbb{R}$, inductively by

$$Y(0,\omega) = y_0$$

$$Y(\underline{t}+\Delta t,\omega) = G(\underline{t},Y(\underline{t}))\Delta X(\underline{t},\omega) + F(\underline{t},Y(\underline{t}))\Delta t.$$

Therefore

(11) $\quad Y(\underline{t}) = y_0 + \int_0^{\underline{t}} G(\underline{s},Y(\underline{s}))dX(\underline{s}) + \int_0^{\underline{t}} F(\underline{s},Y(\underline{s}))d\lambda.$

We claim that Y is S-continuous. The second internal integral is trivially S-continuous by the boundedness of F. The S-continuity of $M(\underline{t}) = \int_0^{\underline{t}} G(\underline{s},Y(\underline{s}))dX(\underline{s})$ is a consequence of Theorem 6 since M is an $(A_{\underline{t}})$-martingale with $[M](\underline{t}) = \int_0^{\underline{t}} G(\underline{s},Y(\underline{s}))^2 d\lambda$. Let $y = st(Y)$. Then y is F_t-adapted and has continuous paths a.s. Let N be the $L(\lambda\times\overline{P})$-null set,

$$N = \{(\underline{t},\omega)\in ns(T)\times\Omega \mid st(\hat{G}(\underline{t})) \neq \hat{\sigma}({}^\circ\underline{t}) \text{ or } st(\hat{F}(\underline{t})) \neq \hat{f}({}^\circ t) \text{ or } Y(\cdot,\omega) \text{ is not S-continuous}\}.$$

If $(\underline{t},\omega)\in(ns(T)\times\Omega) - N$, then using (9) and (10) we have

$${}^\circ G(\underline{t},Y(\underline{t},\omega)) = \sigma({}^\circ\underline{t},{}^\circ Y(\underline{t},\omega)) = \sigma({}^\circ\underline{t},y({}^\circ\underline{t},\omega))$$

$${}^\circ F(\underline{t},Y(\underline{t},\omega)) = f({}^\circ\underline{t},{}^\circ Y(\underline{t},\omega)) = f({}^\circ\underline{t},y({}^\circ\underline{t},\omega)).$$

Therefore $G(\underline{t},Y(\underline{t}))$ is a lifting of $\sigma(t,y(t))$ in $\overline{I}$ (since G is uniformly bounded) and the definition of the Itô integral implies

(12) $\quad st(\int G(\underline{s},Y(\underline{s}))dX(\underline{s}))(t) = \int_0^t \sigma(s,y(s))dB(s) \qquad \forall t\geq 0$ a.s.

Using the nonstandard representation of the Lebesgue integral (1), one gets

(13) $\quad st(\int F(\underline{s},Y(\underline{s}))d\lambda)(t) = \int_0^t f(s,y(s))ds \qquad \forall t\geq 0$ a.s.

(12) and (13) allow us to apply st to both sides of (11) and get

$$y(t) = y_0 + \int_0^t \sigma(s,y(s))dB(s) + \int_0^t f(s,y(s))ds.$$

If B' is another Brownian motion on (Ω,F,P,F_t) then there is an internal transformation ϕ such that $B'(t,\omega) = B(t,\phi(\omega))$ for all $t\geq 0$ a.s. (Theorem 8). It follows from (7) that $y'(t,\omega) = y(t,\phi(\omega))$ is a solution of (8). □

By making only minor changes the above proofs also go through if $\sigma:[0,\infty)\times\mathbb{R}^n\to\mathbb{R}^{n\times d}$ ($n\times d$ matrices), $f:[0,\infty)\times\mathbb{R}^n\to\mathbb{R}^n$, B' is a d-dimensional Brownian motion on (Ω,F,P,F_t), where $\Omega=(\{-1,1\}^d)^T$, and y_0 is an F_0-measurable random variable. Some other possible extensions are as follows:

(a) Let $\sigma:[0,\infty)\times\mathbb{R}^n\times\Omega\to\mathbb{R}^{n\times d}$ and $f:[0,\infty)\times\mathbb{R}^n\times\Omega\to\mathbb{R}^n$ depend on ω, where $\sigma, f(t,x,\cdot)$ are F_t-measurable for all (t,x). (See Keisler [14, Th. 5.14].)

(b) Allow σ and f to depend on the past history of y. That is, $\sigma:[0,\infty)\times C([0,\infty),\mathbb{R}^n)\to\mathbb{R}^{n\times d}$ and $f:[0,\infty)\times C([0,\infty),\mathbb{R}^n)\to\mathbb{R}^n$, where $x|_{[0,t]}=y|_{[0,t]}$ implies $\sigma(t,x)=\sigma(t,y)$ and $f(t,x)=f(t,y)$. Solutions of

$$y(t)=y_0+\int_0^t \sigma(s,y)dB'(s)+\int_0^t f(s,y)ds$$

have been obtained by Hoover and Perkins [12] and Cutland [6](the latter without any continuity condition on f).

(c) Remove the continuity conditions on σ and f in Theorem 9 and assume instead that σ^{-2} (or $(\det\sigma)^{-2}$ in higher dimensions) is bounded (Keisler [14, Th. 10.2]). Although some of Keisler's arguments have been simplified by Lindstrøm [20], the proof still relies on a difficult inequality of Krylov [16] and is therefore less satisfying than the simple and intuitive construction used in the proof of Theorem 9.

The basic idea of the above proof goes back to Skorohod [30, Ch. 3, Sec. 3] but whereas Skorohod must change the underlying space and Brownian motion, B', in order to find a solution, on the above Loeb space one constructs solutions of (8) with respect to any given Brownian motion. (We shall see in Section 6 that this situation remains true for a more general class of Loeb spaces and for equations much more general than (8).) A recent example of Barlow [3] shows that such an existence result is false on a general (Ω,F,P,F_t) even if $0<c\le\sigma(t,x)=\sigma(x)$ and $f=0$.

5. Local Time

The local time of a Brownian path B is

$$\ell(t,x)=d/dx\int_0^t I(B(s)\le x)ds\,,\qquad t\ge 0,\qquad x\in\mathbb{R}.$$

That $\ell(t,x)$ exists and is jointly continuous was shown by Trotter [35]. It is not hard to see that for x fixed, $t\to\ell(t,x)$ is non-decreasing and increases only on $\{t\,|\,B(t)=x\}$. A simple nonstandard construction of local time (independent of Trotter's theorem) is obtained by considering the particular Brownian motion, $B=st(X)$, of section 3 and discretizing the approximation

$$\ell(t,x)\approx\int_0^t I(x\le B(x)\le x+\Delta x)ds(\Delta x)^{-1}.$$

Indeed, replacing B with X, Δx with $(\Delta t)^{1/2}$, ds with $d\lambda$ and assuming that $x\in S=\{k(\Delta t)^{1/2}\,|\,k\in{}^*\mathbb{Z}\}$, we get the following

Definition. The local time of the infinitesimal random walk, X, is the internal process $L: T \times S \times \Omega \to {}^*\mathbb{R}$ defined by

$$L(\underline{t},\underline{x}) = \sum_{0 \le \underline{s} < \underline{t}} I(X(\underline{s}) = \underline{x})(\Delta t)^{\frac{1}{2}}$$

$$= (\text{no. of times } X \text{ visits } \underline{x} \text{ up to time } \underline{t}) \times (\Delta t)^{\frac{1}{2}}. \quad \square$$

Theorem 10 (Perkins [25]). L is S-continuous (on $T \times S$) and $st(L)$ is the local time of B $(= st(X))$. □

Using the nonstandard formulation of weak convergence (see Loeb [22, IV.B, Cor. 3] or Anderson and Rashid [2]) one gets the following result as an immediate corollary:

Corollary 11. Let $\{Y_i\}$ be i.i.d. random variables such that $P(Y_i = \pm 1) = 1/2$ and set $S_n = \sum_{i=1}^n Y_i$. Define $L_n \in C([0,\infty) \times \mathbb{R}, \mathbb{R})$ and $X_n \in C([0,\infty), \mathbb{R})$ by setting

$$L_n(jn^{-1}, kn^{-\frac{1}{2}}) = \sum_{0 \le i < j} I(S_i = k) n^{-\frac{1}{2}}$$

$$X_n(jn^{-1}) = S_j n^{-\frac{1}{2}}$$

for $j \in \mathbb{N}_0$ and $k \in \mathbb{Z}$, and then interpolate polygonally. Then

$$(X_n, L_n) \xrightarrow{w} (B, \ell) \quad \text{on } C([0,\infty), \mathbb{R}) \times C([0,\infty) \times \mathbb{R}, \mathbb{R})$$

($\xrightarrow{w}$ denotes weak convergence). □

The above theorem is proved in [26] along with some more general invariance principles for local time, in which $\{Y_i\}$ is a sequence of i.i.d. random variables of mean zero and variance one.

We now use the nonstandard constructions of the Itô integral and local time to give a simple and intuitive proof of the following formula of Tanaka which describes local time in terms of a stochastic integral:

$$|B(t) - x| = \int_0^t \text{sgn}(B(s) - x)\, dB(s) + \ell(t,x) \quad \forall t \ge 0 \text{ a.s.}, \quad \forall x \in \mathbb{R},$$

where

$$\text{sgn}(x) = \begin{cases} 1 & \text{if } x > 0 \\ 0 & \text{if } x = 0 \\ -1 & \text{if } x < 0 \end{cases}.$$

We give the proof for $x = 0$, as the general case is then obvious. Note that

$$|X(\underline{t} + \Delta t)| - |X(\underline{t})| = \begin{cases} \Delta X(\underline{t}) & \text{if } X(\underline{t}) > 0 \\ (\Delta t)^{\frac{1}{2}} & \text{if } X(\underline{t}) = 0 \\ -\Delta X(t) & \text{if } X(\underline{t}) < 0 \end{cases}.$$

Adding the above increments gives

$$|X(\underline{t})| = \sum_{0 \le \underline{s} < \underline{t}} \text{sgn}(X(\underline{s}))\Delta X(\underline{s}) + \sum_{0 \le \underline{s} < \underline{t}} I(X(\underline{s}) = 0)(\Delta t)^{\frac{1}{2}}$$

(14) $\therefore\ |X(\underline{t})| = \int_0^{\underline{t}} \operatorname{sgn}(X(\underline{s}))dX(\underline{s}) + L(\underline{t},0)$.

Note that

$$\begin{aligned} & L(\lambda\times\overline{P})(\{(\underline{t},\omega)\in ns(T)\times\Omega \mid \operatorname{sgn}(X(\underline{t})) \neq \operatorname{sgn}(B({}^\circ\underline{t}))\}) \\ & \leq \lim_{n\to\infty} L(\lambda\times\overline{P})(\{(\underline{t},\omega) \mid \underline{t}\leq n,\ B({}^\circ\underline{t}) = 0\}) \\ & = \lim_{n\to\infty} E(L(\lambda)(\{\underline{t}\leq n \mid B({}^\circ\underline{t}) = 0\})) \qquad \text{(by Theorem 1)} \\ & = E(\textstyle\int_0^\infty I(B(s) = 0)ds) \qquad \text{(by (0))} \\ & = 0 . \end{aligned}$$

Therefore $\operatorname{sgn}(X(\underline{s}))$ is a lifting of $\operatorname{sgn}(B(s))$ in $\overline{I}$ and taking standard parts in (14) gives

$$|B(t)| = \int_0^t \operatorname{sgn}(B(s))dB(s) + \ell(t,0) \qquad \forall\, t\geq 0 \text{ a.s.},$$

as required.

More importantly, the nonstandard representation of local time has been used to establish some new properties of local time. Notably the following result from [25]:

Theorem 12 (a) Let $m(t,x,\delta)$ denote the Lebesgue measure of the set of points within $\delta/2$ of $\{s\leq t \mid B(s) = x\}$. Then for a.a.$\omega$ and each $T>0$,

$$\lim_{\delta\to 0^+} \sup_{t\leq T,\ x\in\mathbb{R}} |m(t,x,\delta)\delta^{-\frac{1}{2}} - 2(2/\pi)^{\frac{1}{2}}\ell(t,x)| = 0 .$$

(b) Let $m'(t,x,\delta)$ denote the total length of all the excursions of B from x completed by time t, which do not exceed δ in length. Then for a.a.ω and each $T>0$,

$$\lim_{\delta\to 0^+} \sup_{t\leq T,x\in\mathbb{R}} |m'(t,x,\delta)\delta^{-\frac{1}{2}} - (2/\pi)^{\frac{1}{2}}\ell(t,x)| = 0 .$$

(c) Let $n(t,x,\delta)$ denote the number of excursions of B from x completed by time t, which exceed δ in length. Then for a.a.ω and each $T>0$,

$$\lim_{\delta\to 0^+} \sup_{t\leq T,x\in\mathbb{R}} |n(t,x,\delta)\delta^{\frac{1}{2}} - (2/\pi)^{\frac{1}{2}}\ell(t,x)| = 0 . \quad \square$$

The above results were known to hold for a single value of x a.s. (see Kingman [15] and Williams [36,p. 95]). The fact that the uncountable collection of exceptional null sets (one for each x) may be combined into a single null set, off which the convergence is uniform in x, gives 3 characterizations of $\ell(t,x)$ that are both intrinsic (i.e., depend only on $\{s \mid B(s) = x\}$) and global (i.e., hold for all x simultaneously).

Part (a) of the above theorem was in turn used to prove the following result, which answers a question of Taylor and Wendel [34]:

Theorem 13 (Perkins [27]). Let $\phi(t) = (2t|\log|\log t^{-1}||)^{\frac{1}{2}}$. The Hausdorff

ϕ-measure of $\{s \le t \mid B(s) = x\}$ equals $\ell(t,x)$ for all $(t,x) \in [0,\infty) \times \mathbb{R}$ (a.s.). □

(Taylor and Wendel [34] proved the theorem up to a multiplicative constant for $x = 0$.)

The derivation of Theorem 13 from Theorem 12 is completely standard.

6. Stochastic Integration and Stochastic Differential Equations

In this section we study the same problems considered in sections 3 and 4 but in a more general setting. Instead of a Brownian motion, our integrator will be a semimartingale (the precise definition is given below). Two nonstandard approaches to the theory of general stochastic integration were given independently by Lindstrøm ([17],[18],[19]) and Hoover and Perkins ([11],[12]). Due to an obvious bias we outline the latter approach. Proofs are only included when they are trivial, or almost so.

We work for the most part with processes whose sample paths are in

$$D(\mathbb{R}^k) = \{d:[0,\infty) \to \mathbb{R}^k \mid d \text{ right-continuous with left limits}\}.$$

This space is equipped with the Skorohod J_1 topology (see Billingsley [4] and Stone [31]). A nonstandard characterization of this topology (see Proposition 14 below) seems to be simpler than the standard metrics. We alter the definition of our internal time-line, T, so that $T = \{k\Delta t \mid k \in {}^*\mathbb{N} \cup \{0\}\}$, where $\Delta t = 1/\eta$, as before.

<u>Definition</u>. A function $F \in {}^*D(\mathbb{R}^k)$ is SDJ if

(i) $°F(t)$ is finite, whenever $°t < \infty$ $(t \in {}^*[0,\infty))$

(ii) $°F(t) = °F(0)$, whenever $°t = 0$ $(t \in {}^*[0,\infty))$

(iii) For each $t \in [0,\infty)$ there is a point $t_0 \approx t$ such that $F(u) \approx F(t_0^-)$ for all $u \in \{u \approx t, u < t_0\}$ and $F(u) \approx F(t_0)$ for all $u \in \{u \approx t, u \ge t_0\}$.

An internal function $F:T \to {}^*\mathbb{R}^k$ is SDJ if its extension to ${}^*[0,\infty)$ as a step function is SDJ.

If F is SDJ (in either of the above cases), define $st(F):[0,\infty) \to \mathbb{R}^k$ by

$$st(F)(t) = \lim_{°u \to t^+} °F(u),$$

where the limit is over those u in the domain of F such that $°u > t$. □

It is easy to check (Hoover and Perkins [11, Proposition 2.3]) that $st(F)$ is well defined and in $D(\mathbb{R}^k)$. Note also that if F is S-continuous this definition agrees with our earlier one.

The proof of the following nonstandard interpretation of the J_1 topology may be found in Hoover and Perkins [11, Th. 2.6] and Stroyan and Bayod [32].

<u>Proposition 14</u>. $F \in {}^*D(\mathbb{R}^k)$ is SDJ if and only if F is nearstandard in the J_1

topology, and in this case $st(F)$ is the standard part of F in $D(\mathbb{R}^k)$. □

We no longer confine our attention to the particular probability space constructed in section 2 but instead consider a broader class of Loeb spaces. Let $(\Omega, A, \overline{P})$ be an internal probability space and let $\{A_{\underline{t}} | \underline{t} \in T\}$ be an internal filtration, that is, $\{A_{\underline{t}} | \underline{t} \in T\}$ is an internal collection of internal sub-σ-algebras of A such that $A_{\underline{t}} \subseteq A_{\underline{t}+\Delta t}$. Let $(\Omega, F, P) = (\Omega, L(A), L(\overline{P}))$, and define a standard filtration $\{F_t\}$ by

$$F_t = (\bigcap_{{}^\circ\underline{t}>t} \sigma(A_{\underline{t}})) \vee N,$$

where N is the class of P-null sets in F. The 4-tuple (Ω, F, P, F_t) constructed in this manner is called an adapted Loeb space. As before, $\{F_t\}$ is right-continuous (i.e., $F_t = \bigcap_{s>t} F_s$). Definitions and notations from previous sections will be carried over to this more general setting without further comment.

<u>Definitions.</u> A lifting of $\{F_t\}$ is an internal filtration $\{B_{\underline{t}} | \underline{t} \in T\}$ such that

$$F_t = (\bigcap_{{}^\circ\underline{t}>t} \sigma(B_{\underline{t}})) \vee N.$$

An internal process is an internal mapping $X: T \times \Omega \to {}^*Y$ (Y is a topological space equipped with its Borel sets) such that $X(\underline{t}, \cdot)$ is A-measurable for each $\underline{t} \in T$. If $Y = \mathbb{R}^k$, such a process is SDJ if $X(\cdot, \omega)$ is SDJ for almost all ω. In this case, $st(X)$ is the stochastic process, with sample paths in $D(\mathbb{R}^k)$, defined by

$$st(X)(t,\omega) = \begin{cases} st(X(\cdot,\omega))(t) & \text{if } X(\cdot,\omega) \text{ is SDJ} \\ 0 & \text{otherwise} \end{cases} .\quad \square$$

In particular $\{A_{\underline{t}}\}$ is a lifting of $\{F_t\}$. It will, however, often be convenient to work with other liftings. If $x: [0,\infty) \times \Omega \to \mathbb{R}^k$ is a stochastic process with sample paths in $D(\mathbb{R}^k)$ we may view x as a measurable mapping from Ω to the separable metric space $D(\mathbb{R}^k)$. By the lifting theorem (Loeb [22, IV.A, Th. 9]) there is an internal A-measurable mapping $\hat{X}: \Omega \to {}^*D(\mathbb{R}^k)$ such that $st(\hat{X}) = x(\cdot,\omega)$ a.s. If $X: T \times \Omega \to {}^*\mathbb{R}^k$ is defined by $X(\underline{t},\omega) = \hat{X}(\omega)(\underline{t})$, then Proposition 14 shows that X is SDJ and $st(X) = x$ a.s. It is also possible to prove an analogous lifting theorem for (F_t)-adapted processes with sample paths in $D(\mathbb{R}^k)$ (see Hoover and Perkins [11, Th. 4.4]).

<u>Definitions.</u> (a) A stochastic process, x, with sample paths in $D(\mathbb{R}^k)$ is a (k-dimensional) local martingale if there is a sequence of stopping times $\{V_n\}$, increasing to ∞ a.s., such that $x(V_n \wedge t)$ is a uniformly integrable (F_t)-martingale for each n. L^k denotes the set of k-dimensional local martingales.

(b) An (F_t)-adapted process, a, with sample paths in $D(\mathbb{R}^k)$ is a (k-dimensional) process of BV if its sample paths have bounded variation on compacts a.s. V^k denotes the set of k-dimensional processes of BV.

(c) An (F_t)-adapted process, z, with sample paths in $D(\mathbb{R}^k)$ is a (k-dimensional) semimartingale if $z = x + a$ for some $x \in L^k$ and $a \in V^k$.

(d) The σ-field of predictable sets, P, is the σ-field of $[0,\infty) \times \Omega$ generated by all (F_t)-adapted processes whose sample paths are left continuous with right limits. We say that a stochastic process, taking values in some measurable space, is predictable if it is P-measurable. □

Note that the decomposition of z in (c) need not be unique.

Classically one may define the stochastic integral $\int_0^t h(s,\omega)dz(s,\omega)$ for $\mathbb{R}^d$-valued semimartingales, z, and $\mathbb{R}^{n\times d}$-valued (here $\mathbb{R}^{n\times d}$ is the space of $n \times d$ matrices equipped with its usual norm, $|\ |$) predictable integrands, h, which satisfy a boundedness condition. We now show how to extend the nonstandard construction of the Itô integral (section 3) to this more general setting. To avoid messy stopping time arguments we will assume that all our integrands are bounded (the nonstandard construction of the stochastic integral extends easily to the class

$L(z,\mathbb{R}^{n\times d}) = \{h: [0,\infty) \times \Omega \to \mathbb{R}^{n\times d} \mid h$ predictable, and for some $(a,x) \in V^d \times L^d$, $z = a + x$, $(\int_0^t |h(s)|^2 d[m,m]_s)^{1/2}$ is locally integrable and $|h|(\cdot,\omega)$ is integrable with respect to the variation of a, on compacts a.s.$\}$).

The first step (which was immediate in section 3) is to obtain an appropriate lifting of the semimartingale z.

__Definitions.__ Let $\{\mathcal{B}_{\underline{t}} \mid \underline{t} \in T\}$ be a lifting of $\{F_t\}$.

A $(\mathcal{B}_{\underline{t}})$-local martingale lifting of $x \in L^d$ is an SDJ (d-dimensional) $(\mathcal{B}_{\underline{t}})$-martingale, X, satisfying $st(X) = x$ a.s. and for which there is a sequence of $(\mathcal{B}_{\underline{t}})$-stopping times, $\{V_n\}$, such that $X(V_n)$ is S-integrable,

(15) $\quad {}^\circ X(V_n) = x({}^\circ V_n)$ a.s.

and ${}^\circ V_n \uparrow \infty$ a.s. We say that $\{V_n\}$ reduces X.

A $(\mathcal{B}_{\underline{t}})$-BV lifting of $a \in V^{d}$ is a $\mathcal{B}_{\underline{t}}$-adapted SDJ process, A, such that $||A||$ is also SDJ, $st(A) = a$ a.s., and $st(||A||) = ||a||$ a.s. (here $||a||(t)$ is the total variation of a on $[0,t]$ and

$$||A||(\underline{t}) = \textstyle\sum_{\underline{s}<\underline{t}} |\Delta A(\underline{s})|,$$

where $|\ |$ denotes Euclidean distance).

If $x \in L^d$ and $a \in V^d$, a $(\mathcal{B}_{\underline{t}})$-semimartingale lifting of (a;x) is a pair (A;X), where A is a $(\mathcal{B}_{\underline{t}})$-BV lifting of a, X is a $(\mathcal{B}_{\underline{t}})$-local martingale lifting of x, and (A,X) is SDJ (whence A+X is also SDJ). □

Theorem 15 (Hoover and Perkins [11, Th. 7.6]). If $a \in V^d$ and $x \in L^d$, there is a lifting, $\{B_{\underline{t}}\}$, of $\{F_t\}$ and $(B_{\underline{t}})$-semimartingale lifting of $(a;x)$. □

Note that the filtration, $\{F_t\}$, is lifted along with $(a;x)$. If one fixes a lifting, $\{B_{\underline{t}}\}$, in advance it may not be possible to find a $(B_{\underline{t}})$-semimartingale lifting of $(a;x)$ (see Hoover and Perkins [11, Remarks 5.7]). The converse of Theorem 15 is also true (Hoover and Perkins [11, Th. 5.2, Lemma 7.5]). That is, if there is a lifting, $\{B_{\underline{t}}\}$, of $\{F_t\}$, a $(B_{\underline{t}})$-local martingale lifting of x and a $(B_{\underline{t}})$-BV lifting of a, then $x \in L^d$ and $a \in V^d$. (This converse would be false if (15) was omitted in the definition of a $(B_{\underline{t}})$-local martingale lifting.)

Fix a d-dimensional semimartingale, z, and a particular decomposition, $z = a + x$ $(a \in V^d, x \in L^d)$. Choose a lifting, $\{B_{\underline{t}}\}$, of $\{F_t\}$ and a $(B_{\underline{t}})$-semimartingale lifting, $(A;X)$, of $(a;x)$. Let $Z = A + X$. In order to find an appropriate lifting of the predictable process we wish to integrate, the definition of lifting, introduced in section 3, is generalized as follows:

Definition. $\lambda_{(A;X)}(\omega)$ is the internal random measure on T satisfying

$$\lambda_{(A;X)}(\omega)(\{\underline{t}\}) = |\Delta X(\underline{t},\omega)|^2 + |\Delta A(\underline{t},\omega)|$$

and $M_{(A;X)}$ is the (standard) measure on $(T \times \Omega, \sigma(C \times A))$ defined by

$$M_{(A;X)}(B) = E(L(\lambda_{(A;X)}(\omega))(B(\omega) \cap ns(T))),$$

where $B(\omega) = \{\underline{t} \in T \mid (\underline{t},\omega) \in B\}$.

If Y is a Hausdorff space and $h:[0,\infty) \times \Omega \to Y$, then an $(A;X)$-lifting of h is an internal process, $H: T \times \Omega \to {}^*Y$ such that $°H(\underline{t},\omega) = h(°\underline{t},\omega)$ $M_{(A;X)}$ - a.s. □

In order to show that $M_{(A;X)}$ is well-defined one must prove that $\omega \to L(\lambda_{(A;X)}(\omega))(B(\omega) \cap ns(T))$ is $\sigma(A)$-measurable for B in $\sigma(C \times A)$. This follows by first checking it for $B \in C \times A$, and then using a monotone class argument. Note that if X is the infinitesimal random walk of section 3 and $A = 0$, then $\lambda_{(0;X)}(\omega) = \lambda$ and hence

$$M_{(0;X)}(B) = L(\lambda \times \overline{P})(B \cap (ns(T) \times \Omega)).$$

The above definition of lifting therefore generalizes the definition of section 3.

Notation. If $H: T \times \Omega \to {}^*\mathbb{R}^{n \times d}$ is an internal stochastic process, let

$$\int_0^{\underline{t}} H dZ = \sum_{\underline{s}<\underline{t}} H(\underline{s},\omega)\Delta Z(\underline{s},\omega).$$

(Here we consider $\Delta Z(\underline{s},\omega)$ as a column vector.)

Theorem 16 (Hoover and Perkins [11, Th.'s 7.11(a), 7.13(e),(f), 7.16]).

Let Y be a separable metric space and $h:[0,\infty) \times \Omega \to Y$ be predictable. If $\{B_{\underline{t}}'\}$ is any lifting of $\{F_t\}$, then

(a) h has a $(B_{\underline{t}}')$-adapted $(A;X)$-lifting.

(b) If $Y = \mathbb{R}^{n\times d}$ and H is a $(\underline{B_t}')$-adapted $(A;X)$-lifting of h such that

(16) $\quad {}^\circ\sup_{(\underline{t},\omega)} |H(\underline{t},\omega)| < \infty$,

then $\int_0^{\underline{t}} HdZ$ is SDJ and $st(\int HdZ)(t)$ is independent (up to null sets) of the choice of $\{\underline{B_t}'\}$, $(A;X)$ or H (the latter satisfying (16)).

(c) If z also equals $x' + a'$ for some $x' \in L^d$, $a' \in V^d$, and $\int' HdZ$ is defined using this new decomposition then $st(\int HdZ) = st(\int' HdZ)$ a.s. whenever H is a $(\underline{B_t}')$-adapted lifting of h satisfying (16). □

Definition. If $h: [0,\infty) \times \Omega \to \mathbb{R}^{n\times d}$ is uniformly bounded and predictable, choose a $(\underline{B_t})$-adapted $(A;X)$-lifting, H, of h that satisfies (16) and let

$$\int_0^t h(s,\omega)dz(s,\omega) = st(\int HdZ)(t) . \quad \square$$

Theorem 16 shows that the above definition makes sense and (up to null sets) is independent of the choice of $\{\underline{B_t}\}$, $(A;X)$, H or the decomposition $z = a + x$. That the above definition of the stochastic integral agrees with the standard one may be shown by adapting the argument presented for the Itô integral in section 3 (see Hoover and Perkins [11, Th. 7.15(b)]).

Before considering stochastic differential equations, we describe a nonstandard construction of quadratic variation.

Notation. If $n = (n_1,n_2) \in L^2$ and $Q = \{t_0,\ldots,t_L\}$ is a finite partition of $[0,t]$ with $0 = t_0 < \ldots < t_L = t$, let $||Q|| = \max_{1\le i\le L} |t_i - t_{i-1}|$ and

$$S_t(n,Q) = n_1(0)n_2(0) + \sum_{i=1}^{L} (n_1(t_i) - n_1(t_{i-1}))(n_2(t_i) - n_2(t_{i-1})) .$$

Recall the definition of $[Y_1,Y_2]$ given in section 3 for $^*\mathbb{R}$-valued internal processes, Y_1 and Y_2.

Theorem 17 (Hoover and Perkins [11, Th. 6.7, 7.18]). If (N_1,N_2) is a $(\underline{B_t})$-local martingale lifting of $n = (n_1,n_2) \in L^2$, then $[N_1,N_2]$ is SDJ and for each t,

$$\lim_{||Q||\to 0^+} S_t(n,Q) = st([N_1,N_2])(t) ,$$

where the convergence holds in probability. □

Definition. If $(n_1,n_2) \in L^2$ choose a $(\underline{B_t})$-local martingale lifting, (N_1,N_2), of n and let

$$[n_1,n_2](t) = st([N_1,N_2])(t) . \quad \square$$

The above definition is independent of the choice of $\{\underline{B_t}\}$ or (N_1,N_2) and coincides with the standard definition (see Doléans-Dade [7]) by Theorem 17. The nonstandard representations of the stochastic integral and quadratic variation can now be used to give very intuitive proofs of several standard results. A good example is the following integration by parts formula:

Theorem 18. If $n_1, n_2 \in L^1$, then

$$n_1(t)n_2(t) = \int_0^t n_1(s^-)dn_2(s) + \int_0^t n_2(s^-)dn_1(s) + [n_1,n_2](t) \quad \text{a.s.}$$

Proof. Let $\{\mathcal{B}_t\}$ be a lifting of $\{F_t\}$ and (N_1,N_2) be a $(\mathcal{B}_{\underline{t}})$-local martingale lifting of (n_1,n_2) (Theorem 15). Then

$$\Delta(N_1N_2)(\underline{t}) = N_1(\underline{t})\Delta N_2(\underline{t}) + N_2(\underline{t})\Delta N_1(\underline{t}) + (\Delta N_1(\underline{t}))(\Delta N_2(\underline{t})),$$

and therefore

(17) $\quad N_1(\underline{t})N_2(\underline{t}) = \int_0^{\underline{t}} N_1(\underline{s})dN_2(\underline{s}) + \int_0^{\underline{t}} N_2(\underline{s})dN_1(\underline{s}) + [N_1,N_2](\underline{t})$.

We claim that

(18) $\quad N_1$ is a $(0;N_2)$-lifting of $n_1(\bar{s})$.

There is a sequence of $(\mathcal{B}_{\underline{t}})$-stopping times, $\{T_n\}$, such that

$$\{\underline{s}\in ns(T)\,|\,{}^\circ N_1(\underline{s}) \neq {}^\circ N_1(\underline{s}-\Delta t)\} = \{T_n(\omega)\,|\,n\in\mathbb{N}, T_n(\omega)\neq\infty\} \quad \text{a.s.}$$

Since $st(N_1) = n_1$ a.s., it follows that for a.a.ω,

$$\{\underline{s}\in ns(T)\,|\,{}^\circ N_1(\underline{s}) \neq n_1({}^\circ\underline{s}^-)\} \subset \bigcup_{n=1}^{\infty}\bigcap_{m=1}^{\infty}[T_n, T_n + m^{-1}) \equiv \Lambda .$$

The facts that ${}^\circ N_1(T_n) \neq {}^\circ N_1(T_n - \Delta t)$ a.s. on $\{T_n < \infty\}$, $n_2 = st(N_2)$ a.s., and (N_1,N_2) is SDJ imply that ${}^\circ N_2(T_n) = n_2({}^\circ T_n)$ a.s. on $\{T_n < \infty\}$. This shows that $M_{(0;N_2)}(\Lambda) = 0$ and proves (18). By symmetry, N_2 is a $(0;N_1)$-lifting of $n_2(\bar{s})$. The result follows by taking standard parts in (17) and using the definitions of the stochastic integral and $[n_1,n_2]$. □

We are ready to state the existence theorem for stochastic differential equations from Hoover and Perkins [12, Th. 10.3].

Theorem 19. Let $h:[0,\infty)\times\Omega\to\mathbb{R}^n$, $z:[0,\infty)\times\Omega\to\mathbb{R}^n$ and $f:[0,\infty)\times\Omega\times D(\mathbb{R}^n)\to\mathbb{R}^{n\times d}$ satisfy:

(H_1) z is a semimartingale.

(H_2) h is (F_t)-adapted and has sample paths in $D(\mathbb{R}^n)$.

(H_3) f is $P\times\mathcal{D}$-measurable ($\mathcal{D}$ denotes the Borel sets in $D(\mathbb{R}^n)$) and $f(t,\omega,d_1) = f(t,\omega,d_2)$ whenever $d_1|_{[0,t)\cup\{0\}} = d_2|_{[0,t)\cup\{0\}}$.

(H_4) There is a predictable, $\mathbb{R}$-valued process $\bar{f}$, $x\in L^d$, and $a\in V^d$ such that $z = x + a$, $\sup_d |f(t,\omega,d)| \le \bar{f}(t,\omega)$, $\int_0^n \bar{f}(t,\omega)d\|a\|(t) < \infty$ for all $n\in\mathbb{N}$ a.s., and $(\int_0^t \bar{f}(s,\omega)^2 d[m,m](s))^{1/2}$ is locally integrable.

(H_5) $f(t,\omega,\cdot)$ is continuous on $D(\mathbb{R}^n)$ equipped with the topology of uniform convergence on compacts.

Then there is an $(\mathcal{F}_t)$-adapted process, y, with sample paths in $D(\mathbb{R}^n)$ such that

(19) $\quad y(t,\omega) = h(t,\omega) + \int_0^t f(s,\omega,y)dz(s,\omega) \qquad \forall t \geq 0 \quad$ a.s.

Proof. We only give an outline of the proof under the following simplifying assumptions:
(i) $h(t,\omega) \equiv h_0 \in \mathbb{R}^n$.
(ii) $|f|$ is uniformly bounded.
(iii) $f(t,\omega,d) = g(t,\omega,d(t^-))$, where (necessarily) $g(t,\omega,\cdot)$ is continuous on $\mathbb{R}^n$ and $g(\cdot,\cdot,x)$ is predictable $\forall x$.

Only (iii) significantly simplifies the proof.

As above we may choose a lifting, $\{\mathcal{B}_{\underline{t}}\}$, of $\{\mathcal{F}_t\}$ and a $(\mathcal{B}_{\underline{t}})$-semimartingale lifting, $(A;X)$, of $(a;x)$, where $z = a+x$. Let $Z = A+X$. Define $\hat{g}: [0,\infty)\times\Omega \to C(\mathbb{R}^n, \mathbb{R}^{n\times d})$ by $\hat{g}(t,\omega)(x) = g(t,\omega,x)$ and let $\hat{G}$ be a $(\mathcal{B}_{\underline{t}})$-adapted $(A;X)$-lifting of $\hat{g}$ such that ${}^\circ\sup_{(\underline{t},\omega,x)} |\hat{G}(\underline{t},\omega)(x)| < \infty$. Proposition 2 implies that

(20) $\quad M_{(A;X)}({}^\circ\hat{G}(\underline{t},\omega)(x) \neq g({}^\circ\underline{t},\omega,{}^\circ x) \quad \text{for some} \quad x \in ns({}^*\mathbb{R}^n)) = 0$.

Define an internal process, Y, inductively by

$$Y(0,\omega) = h_0$$

$$\Delta Y(\underline{t},\omega) = \hat{G}(\underline{t},\omega)(Y(\underline{t},\omega))\Delta Z(\underline{t},\omega).$$

Therefore

(21) $\quad Y(\underline{t},\omega) = h_0 + \int_0^{\underline{t}} \hat{G}(\underline{s},\omega)(Y(\underline{s},\omega))dZ(\underline{s},\omega)$.

Assume, for the moment, that

(22) $\quad Y$ is SDJ.

Note that

$$|\Delta Y(\underline{t})| > 0 \Rightarrow |\Delta Z(\underline{t})| > 0 \Rightarrow |\Delta(A,X)(\underline{t})| > 0.$$

Since both Y and (A,X) are SDJ, this shows that (Y,A,X) is SDJ. If $y = st(Y)$, then one can now argue, just as in the proof of (18), to see that Y is an $(A;X)$ lifting of $y(t^-)$. This, together with (20), implies that

$${}^\circ\hat{G}(\underline{t},\omega)(Y(\underline{t},\omega)) = g({}^\circ\underline{t},\omega,{}^\circ Y(\underline{t},\omega)) = g({}^\circ\underline{t},\omega,y({}^\circ\underline{t}^-,\omega)) \quad M_{(A;X)}\text{- a.s.}$$

The definition of the stochastic integral allows us to take standard parts in (21) to obtain

$$y(t,\omega) = h_0 + \int_0^t g(s,\omega,y(s^-,\omega))dz(s,\omega),$$

as required.

It remains to show (22). If (a,x) is continuous, then clearly (A,X) is S-continuous and one can use Theorem 6 (just as in the proof of Corollary 7) to prove the following result, which shows that Y is S-continuous:

Theorem 20. If $(A;X)$ is an S-continuous $(\mathcal{B}_{\underline{t}})$-semimartingale lifting of $(a;x)$ and $Z = A + X$, then $\int_0^{\underline{t}} HdZ$ is S-continuous for all bounded, $(\mathcal{B}_{\underline{t}})$-adapted, ${}^*\mathbb{R}^{n\times d}$-valued integrands, H. □

Unfortunately Theorems 6 and 20 both fail if "S-continuous" is replaced by "SDJ", as the following example (due independently to Lindstrøm and Hoover and Perkins [12]):

Example. Let V be an internal geometric random variable such that if $p = (1 + (\Delta t)^{1/2})^{-1}$, then

$$\overline{P}(V = k) = (1-p)\, p^k \qquad k = 0,1,2,\ldots$$

Let $X(\underline{t}) = 0$ for $\underline{t} \le 1$ and

$$\Delta X(1 + j\Delta t) = \begin{cases} (-1)^{j+1}(\Delta t)^{1/2} & \text{if } j < V \\ (-1)^j & \text{if } j = V \\ 0 & \text{if } j > V \end{cases}.$$

Therefore, after $\underline{t} = 1$, X oscillates between $-\sqrt{\Delta t}$ and 0 until V when it has a jump of ± 1. Clearly X is SDJ. An easy computation shows that X is an internal martingale. If $H(1 + j\Delta t) = (-1)^j$ for $1 + j\Delta t \ge 0$, then

$$\int_0^{1+j\Delta t} HdX = \sum_{0 \le i < j\wedge V} -(\Delta t)^{1/2} + I(j > V)$$

$$= \begin{cases} -j(\Delta t)^{1/2} & \text{if } j \le V \\ 1 - V(\Delta t)^{1/2} & \text{if } j > V \end{cases}.$$

Since $V\Delta t \approx 0$ and ${}^\circ(V(\Delta t)^{1/2}) > 0$ a.s., $\int HdX$ is not SDJ.

If we consider X on a coarser time set it will consist (a.s.) of a single jump of ± 1 infinitesimally close to 1 and all the integrals $\int HdX$ (H bounded) become SDJ. In fact something similar will always be true and one can prove the following result (the proof given in Hoover and Perkins [12, Th. 9.7] is long):

Theorem 21. If $(a,x) \in \mathcal{V}^d \times \mathcal{L}^d$, there is a lifting, $\{\mathcal{B}_{\underline{t}}\}$, of $\{\mathcal{F}_t\}$ and a $(\mathcal{B}_{\underline{t}})$-semimartingale lifting, $(A;X)$, of $(a;x)$ such that if $Z = A + X$ then $\int_0^{\underline{t}} HdZ$ is SDJ for all bounded, $(\mathcal{B}_{\underline{t}})$-adapted, ${}^*\mathbb{R}^{n\times d}$-valued integrands, H. □

This result is the heart of the proof and is the main obstacle that distinguishes the general setting from the Itô equations of section 4.

Assume that our liftings, $\{\mathcal{B}_{\underline{t}}\}$ and $(A;X)$, are given by Theorem 21. Then (22) is immediate and the proof of Theorem 19 is complete. □

Remarks.

1. The proof becomes more complicated if (iii) is not assumed since in this case if $\hat{f}(s,\omega)(d) = f(s,\omega,d)$, $\hat{f}$ takes values in the non-separable space $C(D(\mathbb{R}^n), \mathbb{R}^{n\times d})$ and the lifting theorem (Theorem 16(a)) does not apply.

2. Equations of the form (19) were first studied by Doléans-Dade [8] and Protter [28], [29] when $f(t,\omega,d) = g(t,\omega,d(t^-))$ and g satisfies a Lipschitz condition in the space variable. In this case there is a unique solution on an arbitrary adapted probability space (Ω,F,P,F_t).

3. A version of Theorem 19 was proven independently and at about the same time by Jacod and Memin [13] using standard techniques. They show that for a given adapted probability space, (Ω,F,P,F_t) there is a probability measure P' on $(\Omega \times D(\mathbb{R}^n))$, whose projection onto Ω is P, such that $y(\omega,d)(t) = d(t)$ is a solution of (19) on the product space. They then argue that the semimartingale $z'(t,\omega,d) = z(t,\omega)$ on the product space is "the same" as the original semimartingale by showing they have the same local characteristics (the problem here is that the filtration has changed). On an adapted Loeb space, however, solutions of (19) exist with respect to any semimartingale without enlarging the space at all.

4. It is natural to ask if working an adapted Loeb space is at all restrictive. Recent work of Hoover and Keisler [10] indicates that the answer is "no". They show that if $x(t)$ is a stochastic process taking values in a Polish space, M, (i.e., a complete separable metric space) defined on some adapted space $(\mathfrak{X},G,Q,G_t)$, and (Ω,F,P,F_t) is an adapted Loeb space rich enough to support an (F_t)-Brownian motion, then there is a stochastic process $\hat{x}(t)$, defined on (Ω,F,P,F_t) and taking values in M, such that x and $\hat{x}$ have the same adapted distribution. This means that each pair of processes defined from x and $\hat{x}$, respectively, by iterating the two operations

(1) composition with continuous functions

(2) conditional expectation with respect to G_t (respectively, F_t)

have the same finite-dimensional distributions. This is an extremely strong equivalence which seems to preserve all properties of interest of a given process. In fact, Hoover and Keisler [10] show that any adapted Loeb space, (Ω,F,P,F_t), rich enough to support an (F_t)-Brownian motion satisfies the following stronger condition:

If (x,y) is a pair of M-valued processes defined on some $(\mathfrak{X},G,Q,G_t)$ and $\hat{x}$ is defined on (Ω,F,P,F_t) such that x and $\hat{x}$ have the same adapted distribution, then there is a process, $\hat{y}$, on (Ω,F,P,F_t) such that $(\hat{x},\hat{y})$ and (x,y) have the same adapted distribution.

References

1. R.M. Anderson: A non-standard representation for Brownian motion and Itô integration. Israel J. Math. 25, 15-46 (1976).

2. R.M. Anderson and S. Rashid: A non-standard characterization of weak convergence. Proc. Amer. Math. Soc. 69, 327-332 (1978).

3. M.T. Barlow: One dimensional stochastic differential equations with no strong solution, preprint.

4. P. Billingsley: Convergence of probability measures. New York: John Wiley 1968.

5. D.L. Burkholder, B.J. Davis, and R.F. Gundy: Integral inequalities for convex functions of operators on martingales. Proc. 6th Berkeley Symp. 2, 223-240 (1972).

6. N.J. Cutland: On the existence of strong solutions to stochastic differential equations on Loeb spaces, preprint.

7. C. Doléans-Dade: Variation quadratique des martingales continues à droite. Ann. Math. Statist. 40, 284-289 (1969).

8. C. Doléans-Dade: On the existence and unicity of solutions of stochastic integral equations. Z. Wahrscheinlichkeitstheorie verw. Geb. 34, 93-101 (1976).

9. J.L. Doob: Stochastic processes. New York: John Wiley 1953.

10. D.H. Hoover and H.J. Keisler: Adapted probability distributions, preprint.

11. D.H. Hoover and E.A. Perkins: Nonstandard construction of the stochastic integral and applications to stochastic differential equations I, to appear in Trans. Amer. Math. Soc.

12. D.H. Hoover and E.A. Perkins: Nonstandard construction of the stochastic integral and applications to stochastic differential equations II, to appear in Trans. Amer. Math. Soc.

13. J. Jacod and J. Memin: Existence of weak solutions for stochastic differential equations with driving semimartingales. Stochastics 4, 317-337 (1981).

14. H.J. Keisler: An infinitesimal approach to stochastic analysis, to appear as a Memoir of Amer. Math. Soc.

15. J.F.C. Kingman: An intrinsic description of local time. J. London Math. Soc. (2) 6, 725-731 (1973).

16. N.V. Krylov: Some estimates of the probability density of a stochastic integral. Mathematics of the USSR-Izvestia 8, 233-254 (1974).

17. T.L. Lindstrøm: Hyperfinite stochastic integration I: The nonstandard theory. Math. Scand. 46, 265-292 (1980).

18. T.L. Lindstrøm: Hyperfinite stochastic integration II: Comparison with the standard theory. Math. Scand. 46, 293-314 (1980).

19. T.L. Lindstrøm: Hyperfinite stochastic integration III: Hyperfinite representations of standard martingales. Math. Scand. 46, 315-331 (1980).

20. T.L. Lindstrøm: The structure of hyperfinite stochastic integrals, preprint.

21. P.A. Loeb: Conversion from nonstandard to standard measure spaces and applicatio

in probability theory. Trans. Amer. Math. Soc. 211, 113-122 (1975).

22. P.A. Loeb: An introduction to nonstandard analysis and hyperfinite probability theory. Probabilistic Analysis and Related Topics, A. Barucha-Reid Ed., New York: Acad. Press 1979.

23. H.P. McKean: Stochastic Integrals. New York: Academic Press 1969.

24. P.A. Meyer: Un cours sur les intégrales stochastiques. Séminaire de Probabilités X, 245-400. Lect. Notes in Math. 511. Berlin-Heidelberg-New York: Springer 1976.

25. E.A. Perkins: A global intrinsic characterization of Brownian local time. Ann. Probability 9, 800-817 (1981).

26. E.A. Perkins: Weak invariance principles for local time, to appear in Z. Wahrscheinlichkeitstheorie verw. Geb.

27. E. A. Perkins: The exact Hausdorff measure of the level sets of Brownian motion, to appear in Z. Wahrscheinlichkeitstheorie verw. Geb.

28. Ph.E. Protter: On the existence, uniqueness, convergence and explosions of solutions of systems of stochastic integral equations. Ann. Probability 5, 243-261 (1977).

29. Ph.E. Protter: Right-continuous solutions of systems of stochastic integral equations. J. Multivariate Anal. 7, 204-214 (1977).

30. A.V. Skorohod: Studies in the Theory of Random Processes. Reading: Addison-Wesley 1965.

31. C.J. Stone: Weak convergence of stochastic processes defined on semi-infinite time intervals. Proc. Amer. Math. Soc. 14, 694-696 (1963).

32. K.D. Stroyan and J.M. Bayod: Foundations of Infinitesimal Stochastic Analysis, forthcoming book.

33. K.D. Stroyan and W.A.J. Luxemburg: Introduction to the Theory of Infinitesimals. New York: Academic Press 1976.

34. S.J. Taylor and J.G. Wendel: The exact Hausdorff measure of the zero set of a stable process. Z. Wahrscheinlichkeitstheorie verw. Geb. 6, 170-180 (1966).

35. H.F. Trotter: A property of Brownian motion paths. Illinois J. Math. 2, 425-433 (1958).

36. D. Williams: Diffusions, Markov Processes and Martingales. Chichester: John Wiley 1979.

TOWARDS A NONSTANDARD ANALYSIS OF PROGRAMS

M. M. Richter
Lehrstuhl für angewandte Mathematik
Technische Hochschule, Aachen
M. E. Szabo *
Department of Mathematics
Concordia University, Montreal

0. Introduction

The main difficulty in the formal treatment of computer languages is the presence of loops. Such formal considerations arise, for example, in problems connected with program verification, where one introduces logical calculi and proves metatheorems for them. The problem presented by the loops stems from the fact that they represent in some sense infinitary objects. To overcome this difficulty one usually proceeds in two ways: (1) One restricts to finitary iterations. Here all functions are still total. (2) The loops are introduced as certain limit objects, e.g., in some function spaces. Here one also gets partial functions. The nonstandard approach introduces the limit objects as ideal points and keeps functions total. The main classical results in this theory are due to D. Scott (cf. SCOTT and STRACHEY (1971)). One way of using Scott's theory is to assign a semantics to a programming language by translating the commands into terms of the lambda calculus and then interpreting those terms in a model of that calculus. In this paper we will interpret the commands

* The research of the second author was supported in part by Grant No. A8224 of the Natural Sciences and Engineering Research Council of Canada and by a Grant from the Department of Education of Quebec.

directly in a nonstandard model of Peano arithmetic. This allows us to avoid loops and to keep all computations as well as all the corresponding logical calculi formally finite. As will be pointed out below, this is closely connected with Scott's approach (and in some sense just the nonstandard version of it); however, our method enjoys the usual intuitive advantages of the nonstandard analysis of limit processes.

1. *Nonstandard programs*

Our nonstandard theory and terminology are based on the axiomatic approach developed in NELSON (1977). Here the nonstandard world is presented in terms of the usual Zermelo-Fraenkel set theory with choice, augmented by an extra predicate, *standard* (x). Set-theoretical formulas involving this predicate are called *external* and all other formulas are called *internal*. In our development, we require the following concepts based on this distinction:

1.1. Definition (Semi-internal and topological formulas).

(1) A formula Φ is *semi-internal* if it is of the form $(\forall^{st} x \in y)\varphi$, where φ is internal and where the quantifier $\forall^{st}$ ranges only over standard objects.

(2) A formula is called *topological* if the predicate standard(x) occurs only negatively and if the quantification over such variables x is bounded by some set.

On the basis of Nelson's axioms, we have:

1.2. Theorem (Equivalence theorem for topological formulas).

(1) The semi-internal and the topological formulas are equivalent.

(2) If a formula $\varphi(x)$ defines a non-empty set X, then $\varphi(x)$ is equivalent to a topological formula if and only if X is the monad of a standard filter F.

Proof. Cf. RICHTER (1981). []

For the sake of simplicity, we begin our discussion with an analysis of the language of register programs. The semantics of such programs is given by register machines mapping n-tuples of natural numbers partially to n-tuples of natural numbers. We then introduce a verification language for such programs and give an example of a program verification in a sequent calculus formulated in this language. We are particularly interested in contrasting the treatment of loops in the standard and nonstandard versions of this calculus. For a detailed formal description of the syntax involved, we refer to KRÖGER (1976, 1977) and SZABO (1980, 1981).

Our nonstandard n-register programs are defined by induction on the structure of these programs. We let $n \in N$ be a fixed natural number and let i and k range over the elements of a fixed nonstandard extension $*N$ of N.

1.3. Definition (Register programs).

(1) The elementary programs are A_i^n, S_i^n, and I_n^n, with $1 \leq i \leq n$.

(2) If r and s are programs, then so are $r \circ s$, $(r)_i^k$, $(r)_i$, and $(r)^{(i)}$.

(3) The n-register programs are obtained from (1) by finitely, possibly hyperfinitely many, applications of (2).

The intended meaning of these programs is as follows:

(1) A_i^n denotes *addition*. It represents the instruction of adding 1 to the contents of register i.

(2) S_i^n denotes *subtraction*. It represents the instruction of subtracting 1 from the contents of register i if that contents is non-zero and of keeping zero otherwise.

(3) I_n^n denotes the *identity* instruction.

(4) $r \circ s$ denotes the *composition* instruction.

(5) $(r)_i^k$ denotes *bounded conditional iteration*. It represents

the instruction of repeating r until register i contains a zero, but iterating at most k times.

(6) $(r)_i$ denotes *unbounded conditional iteration*. It represents the instruction or repeating r until register i contains a zero.

(7) $(r)^{(i)}$ denotes *bounded unconditional iteration*. It represents the instruction of repeating r as often as the number contained in register i. It is assumed that r does not affect the contents of this register.

Programs without unbounded iteration will play a special rôle in our considerations below. We shall refer to them as *bounded programs*.

For descriptive purposes, it is often convenient to identify programs with their semantics, the machines, and to visualize the effect of a program in terms of its computations. Every machine r processes an input $\vec{x} = \vec{x}_1 = < x_1,\ldots,x_n > \in ({}^*N)^n$ in discrete steps. This action defines a sequence

$$\vec{x}(r) = < \vec{x}_1(r),\ldots,\vec{x}_k(r), \vec{x}_{k+1}(r),\ldots>$$

called a *run* of the program. For unbounded machines, runs may be infinite. We observe, however, that the terms $\vec{x}_k(r)$ and $\vec{x}_{k+1}(r)$ differ in at most one register by at most 1. As the work of CZIRMAZ (1981) and ANDRÉKA, NÉMETI, and SAIN (1980) shows, infinite runs are intimately connected with the adequacy of certain verification procedures for programs. We shall return to this point later.

Machines may be nonstandard for two reasons: (1) They may contain bounded conditional iterations $(r)_i^k$ for non-standard numbers k. (2) They may be built up by means of hyperfinitely many compositions. However, standard machines always yield standard runs for standard inputs.

The fundamental semantical connection between n-register machines is captured in the following definition:

1.4. Definition (Equivalence of programs).

(1) $r \sim_{\vec{x}} s$ if and only if $\vec{x}_k(r) = \vec{x}_k(s)$ for all standard k (i.e., if and only if r and s compute the same values for $\vec{x}$ at all standard times).

(2) $r \sim s$ if and only if $r \sim_{\vec{x}} s$ for all standard $\vec{x}$.

The external relation "$\sim$" has the properties of an equivalence relation. We shall see that this relation makes unbounded machines unnecessary.

1.5. Theorem (Upper bound theorem).

For each standard program r there exists a bounded program s such that $r \sim s$.

Proof. Let R_n be the set of n-register machines and choose a nonstandard natural number ω and define the following nonstandard syntactic mapping

$$B = B(n,\omega): R_n \to R_n$$

on the construction of programs:

(1) $B(r) = r$ if r is an elementary program.

(2) $B(r \circ s) = B(r) \circ B(s)$.

(3) $B((r)_i^k) = (B(r))_i^k$.

(4) $B((r)^{(i)}) = (B(r))^{(i)}$.

(5) $B((r)_i) = (B(r))_i^{\omega}$.

An induction shows that $s = B(r)$ has the desired properties for each standard r. []

In order to invert the mapping B we introduce the *length* $\ell(r)$ of a program r by putting

(1) $\ell(r) = 1$ for elementary programs r.

(2) $\ell(r \circ s) = \ell(r) + \ell(s)$.

(3) $\ell((r)_i^k) = \ell((r)_i) = \ell((r)^{(i)}) = \ell(r) + 1$.

1.6. Theorem (Termination theorem).
For each program r with standard length there exists a standard program s such that $r \sim s$.

Proof. We modify the mapping B in the proof of Theorem 1.5 as follows:

$$(3')\ B((r)_i^k) = (r)_i^k \text{ if } k \text{ is standard.}$$
$$= (r)_i \text{ if } k \text{ is nonstandard.}$$
$$(4')\ B((r)_i) = (B(r))_{(i)}.$$

The axioms for standard sets guarantee that for each r with standard length, the program $B(r)$ is standard. As before, $s = B(r)$ is the desired program. []

In the present context, the value of Theorem 1.6 lies in the fact that we may assume that all computations terminate: some at standard times and some at nonstandard times.

Our verification language for register programs is a nonstandard modification of the Kröger-type temporal language described in KRÖGER (1976,1977), augmented with iteration variables, hyperfinite iterations, and hyperfinite disjunctions. Although this language contains loop formulas, Theorems 1.5 and 1.6 allow us to dispense with such formulas in the associated verification calculus. Our notation is essentially that of SZABO (1980).

We distinguish between *individual variables* $x_0, x_1, \ldots$ (usually written as $x, y, \ldots$), with x_i ranging over the contents of register i and *iteration variables* $n_0, n_1, \ldots$ (usually written as $i, j, k, m, n, \ldots$) ranging over our nonstandard time scale $*N$. We think of the elements of $*N$ as discrete points in time at which computations take place. The *terms* and *formulas* of our language are defined inductively:

1.7. Definition (Verifying terms).

(1) Every individual variable is a term.

(2) The elements of $*N$ are terms.

(3) If t is a term, then so are $(t + 1)$ and $(t - 1)$.

(4) Terms are obtained from (1) and (2) by finitely many applications of (3).

1.8. *Definition* (Verifying formulas).

(1) For each non-zero element α of $*N$, $I(\alpha)$ is a formula.

(2) If t and s are terms, then $(t = s)$ and $(t < s)$ are formulas.

(3) If x is an individual variable and t is a term, then $[x:=t]$ is a formula.

(4) If α is an element of $*N$ and A is a formula, then $(\alpha)A$ is a formula.

(5) If n is an iteration variable or an element of $*N$, then A^n is a formula.

(6) If $A(m)$ is a formula containing the iteration variable m and k is an element of $*N$, then $\bigvee_{0 \leq m \leq k} A(m)$ is a formula

(7) If A and B are formulas, then so are $(-A)$ and $(A \vee B)$.

(8) If A is a formula, then $@A$ is a formula.

(9) Formulas are obtained from (1) - (3) by finitely many applications of (4) - (7).

The truth values (0 and 1) of the formulas defined in 1.8 are thought of as varying along the time scale $*N$ and a formula is *true* if and only if it is true in all models

$$M = \langle M_\sigma \rangle_{\sigma \in *N},$$

where a formula A is true in M if $M_\sigma(A) = 1$ for all $\sigma \in *N$. The definition of the truth of the formula A at time σ requires that we have a measure of the temporal complexity of A. We call this measure the *time dependence* of A and denote it by $\tau(A)$. For loop-free formulas, this measure is independent of the models and is defined as follows:

1.9. Definition (Time dependence of loop-free formulas).

(1) $\tau(t = s) = \tau(t < s) = \tau([x:=t]) = 1$.

(2) $\tau(I(\alpha)) = \alpha$.

(3) $\tau((\alpha)A) = \alpha + \tau(A)$.

(4) $\tau(A^n) = \tau(A) \times n$.

(5) $\tau(\bigvee_{0 \leq m \leq k} A(m)) = \tau(A(k))$.

(6) $\tau(-A) = \tau(A)$.

(7) $\tau(A \vee B) = \max(\tau(A), \tau(B))$.

In the nonstandard context, the time dependence of a loop formula @A varies with our models. There are three reasons: (1) A loop formula is essentially an existential iteration statement and as such should depend on the assignment of constants to the iteration variables. (2) In contrast to the the ordinal time scale $[0,\omega^{\omega})$ used in KRÖGER (1976, 1977), our nonstandard scale $^*N$ has no smallest infinite element which can be used as an upper bound for finite iterations. (3) We allow hyperfinitely many iterations. We therefore, assume that each mapping M_σ assigns to the sequence $< @, \ldots, @, \ldots >$ of copies of @ a sequence $< i_0, \ldots, i_n, \ldots >$ of elements $^*N$, with i_n ambiguously denoted by $M_\sigma(@)$, counting the number of iterations of the formula A in the interpretation of the formula @A at time σ . Here we are assuming that the outermost loop symbol in @A is introduced at the i-th stage in the construction of this formula from simpler formulas. We can now specify the remaining properties of the mapping M_σ:

1.10. Definition (Semantics of verifying formulas).

(1) $M_\sigma(x_i) \in {}^*N$ for each individual variable x_i.

(2) $M_\sigma(n) = n \in {}^*N$ for all terms $n \in {}^*N$.

(3) $M_\sigma(t + 1) = M_\sigma(t) + 1$.

(4) $M_\sigma(t - 1) = M_\sigma(t) - 1$.

(5) $M_\sigma(n_i) \in {}^*N$ for each iteration variable n_i.

(6) $M_\sigma(I(\alpha)) \in \{0,1\}$, subject to two compatibility conditions: If the formula $I(\alpha)$ receives the value 1, then the mappings M_σ and $M_{\sigma+\alpha}$ coincide and $M_\rho(I(\beta)) = 1$ for all $\sigma \leq \rho \leq \rho + \beta \leq \sigma + \alpha$.

This leads to the following definition of the truth values of verifying formulas at time σ:

1.11. Definition (Truth values of verifying formulas).

(1) $M_\sigma(t = s) = 1$ if and only if $M_\sigma(t) = M_\sigma(s)$.

(2) $M_\sigma(t < s) = 1$ if and only if $M_\sigma(t) < M_\sigma(s)$.

(3) $M_\sigma([x:=t]) = 1$ if and only if $M_{\sigma+1}(x) = M_\sigma(t)$.

(4) $M_\sigma((\alpha)A) = 1$ if and only if $M_{\sigma+\alpha}(A) = 1$.

(5) $M_\sigma(A^0) = 1$ if and only if $M_\sigma(I(1)) = 1$.

(6) $M_\sigma(A^n) = 1$ if and only if $M_{\sigma+\tau(A)\times m}(A) = 1$ for all $0 \leq m < M_\sigma(n)$.

(7) $M_\sigma(\bigvee_{0\leq m\leq k} A(m)) = 1$ if and only if $M_\sigma(A(M_\sigma(m))) = 1$ for some numerical instance $A(M_\sigma(m))$ of $A(m)$, with $0 \leq M_\sigma(m) \leq k$.

(8) $M_\sigma(-A) = 1$ if and only if $M_\sigma(A) = 0$.

(9) $M_\sigma(A \vee B) = 1$ if and only if $M_\sigma(A) = 1$ or $M_\sigma(B) = 1$.

(10) $M_\sigma(@A) = 1$ if and only if $M_{\sigma+\tau(A)\times m}(A) = 1$ for all numbers m, $0 \leq m < M_\sigma(@)$ and $M_{\sigma+\tau(A)\times M_\sigma(@)}(I(\tau(A)\times\nu)) = 1$ for some $\nu \in {}^*N - N$, with $M_\sigma(@) < \nu$.

From now on, we shall use customary abbreviations such as $(0 \leq i \leq n)$, $(A \wedge B)$, and $(A \Rightarrow B)$ for verifying formulas whenever convenient. In addition, we write

$$(A \Delta B) \quad \text{and} \quad (A \sim B)$$

in place of $(A \wedge \tau(A)B)$ and $(A \Rightarrow \tau(A)B)$, with the intended meaning of "A and later B" and "If A then later B", where "later" means the time dependence of A.

We are now in a position to define a faithful mapping K from our n-register programs to the given verification language and show how the correctness assertions of some simple programs can be verified by valid formulas, derived in a sequent calculus adapted from SZABO (1980). For a precise specification of the rules involved, we refer to that paper. In each example it can easily be checked on the basis of the given semantics, that the instances of the rules involved take valid formulas to valid formulas and that the axioms occurring in the derivations are valid. In the present context, our objective is to compare the rules of inference governing the loop and the bounded disjunction operators.

The mapping K is defined by an induction on the construction of the n-register programs:

1.12. Definition (Verifying formulas for register programs).

(1) $K(A^n_i) = (0 \leqslant i \leqslant n) \wedge [x_i := x_i + 1]$.

(2) $K(S^n_i) = (0 \leqslant i \leqslant n \wedge [x_i := x_i - 1]$.

(3) $K(I^n_n) = [x_0 := x_0] \Delta \ldots \Delta [x_n := x_n]$.

(4) $K(r \circ s) = K(s) \Delta K(r)$.

(5) $K((r)^{(i)}) = K(r)^{(i)}$.

(6) $K((r)_i) = @((0 < x_i) \wedge K(r)) \Delta ((0 = x_i) \wedge I(1))$.

(7) $K((r)^k_i) = \bigvee_{0 \leqslant m \leqslant k} ((0 < x_i) \wedge K(r))^m \Delta (((0 = x_i) \vee (m = k)) \wedge I(1))$.

The correctness of the elementary program A^n_i, for example, is expressed by the formula

$$((x_i = t) \wedge K(A^n_i)) \sim (x_i = t + 1),$$

which has the following Gentzen-type derivation:

$$\begin{array}{c}
x = t \to x + 1 = t + 1 \qquad [x:=x + 1] \to [x:=x + 1] \\
\hline
x = t,\ [x:=x + 1] \ \to\ [x:=x + 1] \wedge (x + 1 = t + 1) \\
\hline
x = t,\ [x:=x + 1] \ \to\ (1)(x = t + 1) \\
\hline
x = t,\ 1 \leqslant i \leqslant n,\ [x:=x + 1] \ \to\ (1)(x = t + 1) \\
\hline
(x = t) \wedge (1 \leqslant i \leqslant n) \wedge [x:=x + 1] \ \to\ (1)(x = t + 1) \\
\hline
\to\ ((x = t) \wedge K(A^n_i)) \sim (x = t + 1)
\end{array}$$

with x in place of x_i and "→" denoting the Gentzen arrow.

Next we look at a program whose standard verifying formula involves a loop. For the sake of simplicity, we consider the program $(S_i^n)_i$, which reduces the contents of the i-th register to zero. The correctness of this program is expressed by the formula

$$((0 < x - 1) \wedge @((0 < x - 1) \wedge [x := x - 1]) \Delta ((0 = x - 1) \wedge I(1))) \rightsquigarrow (0 = x)$$

with x in place of x_i. Let Φ denote this formula. We sketch a derivation of Φ. The standard version of this derivation involves an instance of the loop rule

$$\frac{\Gamma, (\alpha)(A^n \Delta I(\tau(A) \times \omega), \Delta \rightarrow \Theta \quad \text{for all } n \in N}{\Gamma, (\alpha)(@A), \Delta \rightarrow \Theta}$$

as formulated in KRÖGER (1977), whose infinitely many premises are proved by induction. We first derive

$$0 < x - 1,\ A^n \Delta I(\omega),\ (\omega)(0 = x - 1 \wedge I(1)) \rightarrow (\omega + 1)(0 = x)$$

by induction on n, where $A = (0 = x - 1 \wedge [x := x - 1])$.

Induction Basis:

$$\begin{array}{c}
0<x-1, 0=x-1 \rightarrow \\ \hline
0<x-1, I(1), (1)(0=x-1) \rightarrow \\ \hline
0<x-1, I(1), (1)(0=x-1) \rightarrow (1)(0=x) \\ \hline
0<x-1, I(1), (1)(0=x-1), (1)I(1) \rightarrow (1)(1)(0=x) \\ \hline
0<x-1, I(1), (1)I(\omega), (1)(\omega)(0=x-1), (1)(\omega)I(1) \rightarrow (1)(\omega)(1)(0=x) \\ \hline
0<x-1,\ A^0 \Delta I(\omega), (\omega)(0=x-1 \wedge I(1)) \rightarrow (\omega+1)(0=x)
\end{array} \qquad (0)$$

$$\begin{array}{c}
0<x-1, 0=x \rightarrow \\ \hline
0<x-1, [x:=x-1], (1)(0=x-1) \rightarrow \\ \hline
0<x-1, 0<x-1, [x:=x-1], (1)(0=x-1) \rightarrow (1)(0=x) \\ \hline
0<x-1, A, (1)(0=x-1), (1)I(1) \rightarrow (1)(1)(0=x) \\ \hline
0<x-1, A, (1)I(\omega), (1)(\omega)(0=x-1), (1)(\omega)I(1) \rightarrow (1)(\omega)(1)(0=x) \\ \hline
0<x-1, A \Delta I(\omega), (\omega)(0=x-1 \wedge I(1)) \rightarrow (\omega+1)(0=x)
\end{array} \qquad (1)$$

Induction step:

$$\begin{array}{c}
0<x-1, A^n, (n)I(\omega), (n)B \rightarrow (n)C \\ \hline
0<x-1, A^n, (n)(0<x-1), (n)[x:=x-1], (n+1)I(\omega), (n+1)B \rightarrow (n+1)C \\ \hline
0<x-1, A^{n+1} \Delta I(\omega), B \rightarrow C
\end{array} \qquad (2)$$

where $B = (\omega)(0 = x - 1 \wedge I(1))$ and $C = (\omega + 1)(0 = x)$.

Using these formulas as premises of the loop rule, we derive the formula Φ as follows:

$$\frac{\dfrac{\dfrac{\text{For all } n \in N, \quad 0<x-1, A^n \Delta I(\omega), (\omega)(0<x-1 \wedge I(1)) \to (\omega+1)(0=x)}{0<x-1 \wedge @A, (\omega)(0<x-1 \wedge I(1)) \to (\omega+1)(0=x)}}{(0<x-1) \wedge @A \,\Delta\, (0<x-1 \wedge I(1)) \to (\omega+1)(0=x)}}{\to \Phi} \qquad (3)$$

In our nonstandard world, Derivation (3) makes no sense since the time dependence of $@A$ is not determined absolutely and since the operations Δ and $\frown$ are therefore not defined. We overcome this difficulty by invoking Theorems 1.5 and 1.6 which guarantee that our program $(S_i^n)_i$ is equivalent to a bounded program $(S_i^n)_i^k$, for some infinite natural number k. Hence we can use the bounded disjunction rule

$$\frac{\Gamma, (\alpha)A^n, \Delta \to \Theta \quad \text{for all} \quad n \in [0,k]}{\Gamma, (\alpha)\bigvee_{0\leq m\leq k} A(m), \Delta \to \Theta}$$

in place of the standard loop rule, and prove the modified correctness formula

$$((0 < x - 1) \wedge K((S_i^n)_i^k)) \frown (0 = x)$$

by applying the Transfer Principle to the premisses of Derivation (3) and constructing the obvious variant of Derivation (3). A detailed discussion of the semantic questions involved may be found in SZABO (1981).

By passing to a nonstandard programming theory, we are thus able to replace the infinitary loop rule by the hyperfinite bounded disjunction rule. To obtain a truly finitary verification calculus such as that of Floyd-Hoare, however, one has to reduce the number of valid programs (and thus the number of verifiable programs) by generalizing the notion of model to include a more general class of nonstandard runs of programs. This approach is taken in CZIRMAZ (1981) and in ANDRÉKA, NÉMETI, and SAIN (1980).

2. *Nonstandard programs with procedures*

In the usual programming languages our register programs are expressed more efficiently by instructions such as "While φ do r" for unbounded iteration and "If φ do r else s" for the distinction of cases. Here r and s are programs and φ is a condition (e.g., that the contents of a certain register is non-zero). But there are also programming concepts that are not covered by our simple register machine programming language. An example is the following ALGOL-60 program:

```
Integer procedure P(x);
Integer x;
Begin  P:= If  x > 100 then x - 10 else  P(P(x + 11));
End.
```

Here we have a recursive definition referring to an unspecified procedure P . The intuitive meaning of P is of course clear: We go through the loop until we escape it; if the latter is impossible, the procedure is undefined. In order to be able to write this kind of program in our language, we enlarge the basic vocabulary by means of a new countable (standard) set whose elements serve as *procedure variables* and generalize our class of register programs accordingly. For algebraic reasons, we shall usually refer to these generalized register programs as *procedure terms*.

2.1. Definition (Procedure terms).

(1) Every procedure variable is a procedure term.

(2) Every elementary program is a procedure term.

(3) If r and s are procedure terms, then so are $r \circ s$, $(r)_i^k$, $(r)^{(i)}$, and $(r)_i$.

(4) Every procedure term is obtained from (1) and (2) by finitely many, possibly hyperfinitely many, applications of (3).

The procedure terms form a free algebra (the *term algebra*) generated by the procedure variables. This means that each mapping σ from the procedure variables to the procedure terms has a unique homomorphic extension (also denoted by σ) to the term algebra. This type of homomorphism is usually called a *substitution*. We refer to σ as *variable-free* with respect to a term t if and only if $\sigma(t)$ contains no procedure variable. Since variable-free substitutions convert procedure terms into "real" programs, we can think of procedure terms as "potential" programs.

We compare procedure terms as follows:

2.2. *Definition* (Equivalence of procedure terms).

(1) $t \approx t'$ if and only if $\sigma(t) \sim \sigma(t')$ for every variable-free σ.

(2) $t \leqslant t'$ if and only if $\sigma(t)$ is a subterm of t' for some σ.

(3) $t < t'$ if and only if $t \approx r \leqslant s \approx t'$ for some terms r and s.

The next theorem is basic for our semantics of generalized register programs:

2.3. *Theorem* (Fixed point theorem).

For all standard procedure terms t and all procedure variables f there exists a substitution σ such that $t(s) \approx s$ (where $s = \sigma(f)$ and $t(s)$ is $\sigma(t)$). Furthermore, we have $s < s'$ for all s' with the property that $\sigma'(t) \approx s'$ (where $\sigma'(f) = s'$ and σ' keeps all other variables unchanged).

Proof. For a given standard term t we first define a substitution σ acting as the identity on all variables except f. If f does not occur in t, we put $\sigma(f) = t$. Otherwise we define a substitution τ by $\tau(f) = \tau$, and put $\tau^1 = \tau$ and $\tau^{n+1} = \tau \circ \tau^n$, and take $\sigma = \tau^\omega$ for some nonstandard natural number ω. In the first case we have $\sigma(t) = t = s = \sigma(f)$ and therefore $\sigma(t) \approx s$ holds trivially. In the second case, where f occurs in t, we have $\sigma(f) = \tau^{\omega-1}(f)$ and where the equivalence $\tau^{\omega}(t) \approx \tau^{-1}(t)$ has to be verified. For this purpose, we show that $\eta^{\omega}(t) \approx \eta^{\omega-1}(t)$ for every substitution η such that $\eta(g) = g$ for all $g \neq f$. We proceed by an induction on the structure of the term t. If t is an elementary program or a program variable different from f, we get equality. For the variable f itself, induction shows that for all standard n and for all variable-free substitutions ξ, the runs of $\xi(\eta^{n+1}(f))$ and $\xi(\eta^{n}(f))$ coincide at least up to time n for each standard input n. For the remaining steps in the structural induction on t we use the fact that substitutions are homomorphisms and that "$\sim$" behaves like a congruence relation.

Finally suppose that there is some other σ' such that $\sigma'(f) = s'$, with s' standard, such that $t(s') \approx s'$ holds. If f does not occur in t or if $f = t$, we obtain $\sigma'(t) = t \approx s'$. Hence we assume that $t \neq f$ and f occurs in t. By Theorem 1.5 we may assume that all iterations in s and s' are bounded. Furthermore, we may assume without loss of generality that the same ω occurs in all nonstandard iterations $(r)_i^{\omega}$ and that $t' \approx t$ holds for no proper subterm t' of t. Consider now an arbitrary variable-free substitution η applied to $t(s')$. Then either only subterms of s' are used for arbitrary runs of standard inputs. In that case we put $\xi(f) = s'$ and get $s \leqslant \xi(s) \approx s'$ and hence $s < s'$. Or we have the case that parts of t are also used. Then these parts occur in s' infinitely often and we get $s \approx t(s) \approx t(s') \approx s'$. In this last case, $t(s)$ does not depend (up to equivalence) on what is substituted for f. []

If we apply the transformation described in this proof to our ALGOL-60 procedure mentioned above, we obtain (up to equivalence) the procedure

<u>If</u> $x > 100$ <u>then</u> $x - 10$ <u>else</u> 91 .

Our different transformations on terms t can be summarized as follows:

(1) Use transformation B to eliminate all unbounded iterations from t .

(2) Apply a substitution to t which identifies all procedure variables occurring in t .

(3) Use the fixed point theorem to convert t to its minimum fixed point.

(4) Replace all procedure variables in t by the program $(I)^{\omega}$, with ω nonstandard.

Let $\hat{t}$ be the term obtained from t by means of these transformations. Since $\hat{t}$ is an ordinary bounded register program, we can now define the semantics of generalized register programs in terms of the semantics of their transforms:

2.4. Definition (Semantics of generalized register programs). The semantics of a generalized register program t is the register machine belonging to the bounded register program $\hat{t}$.

The required extension of our verification language and of the associated verification calculus is obtained by including the procedure variables as new (atomic) formulas in our language. The relevant standard systems in KRÖGER (1976, 1977) are already designed in this way.

3. *Generalizations*

The technique described so far can be generalized in various ways. Instead of natural numbers, arbitrary domains A with more general "elementary functions" can be used. We can also introduce functionals of higher type. We indicate this possibility by discussing unary functionals.

3.1. Definition (Computable functionals).

(1) A functional $F : A^A \to A^A$ is *computable* if and only if

$f \approx g$ implies $F(f) \approx F(g)$ for all f, g.

(2) If $F, G : A^A \to A^A$, then $F \approx G$ holds if and only if $F(f) \approx G(f)$ for computable f.

This definition can easily be extended to functionals of arbitrary finite type. Intuitively speaking, a computable functional preserves the standard part of a computation (the "really existing part") step by step. One can also change this feature by not requiring a step-by-step preservation, but allowing to sum up standard finitely many steps.

We now point out the relationship between the nonstandard approach with the classical theory. It is immediate that the definitions of "$\sim$" and "$\approx$" represent topological formulas. Hence these relations are (external) monads of standard filters. The filters are in fact uniformities because the relations are actually equivalence relations. The topologies obtained are the well-known Scott topologies. In these topologies, the set $\{ r_j \mid j \in J \}$ is an approximation of r if the runs of the r_j and r coincide on larger and larger initial segments, and computable functionals are nothing but continuous functions. It is also possible to use these methods to obtain Scott's limit model for the lambda calculus by an (external) factorization of the nonstandard model.

4. *References*

H. ANDRÉKA, I. NÉMETI, and I. SAIN (1980), *Nonstandard runs of Floyd-provable programs*, Proceedings of the Symposium on Algorithmic Logic, A. Salwicki (editor), Lecture Notes in Computer Science, Springer-Verlag, to appear.

L. CZIRMAZ (1981), *Completeness of Floyd-Hoare program verification*, Theoretical Computer Science, 2, pp. 199-211.

C. A. R. HOARE (1969), *An axiomatic basis for computer programming*, Communications of the Association for Computing Machinery, *12*, pp. 576-580, 583.

C. A. R. HOARE (1971), *Procedures and parameters: An axiomatic approach*, Symposium on Semantics of Programming Languages, E. Engeler (editor), Lecture Notes in Mathematics, *188*, Springer-Verlag, New York, pp. 102-116.

F. KRÖGER (1976), *Logical rules of natural reasoning about programs*, Automata, Languages, and Programming, S. Michaelson and R. Milner (editors), Edinburgh University Press, Edinburgh, pp. 87-98.

F. KRÖGER (1977), *LAR: A logic for algorithmic reasoning*, Acta Informatica, *8*, pp. 242-266.

E. NELSON (1977), *Internal set theory*, Bulletin of the American Mathematical Society, *83*, pp. 1165-1198.

M. M. RICHTER (1981), *Monaden, ideale Punkte und Nichtstandard-Methoden*, Vieweg-Verlag, 1981.

D. SCOTT and C. STRACHEY (1971), *Towards a mathematical semantics for computer languages*, Technical Monographs PRG-6, Oxford University Computing Laboratory, Programming Research Group.

D. SCOTT (1974), *Data types as lattices*, Proceedings of the International Summer Institute and Logic Colloquium, Kiel 1974, G. H. Müller, A. Oberschelp, and K. Pothoff (editors), Lecture Notes in Mathematics, *499*, pp. 579-651.

M. E. SZABO (1980), *A sequent calculus for Kröger logic*, Proceedings of the Symposium on Algorithmic Logic, A. Salwicki (editor), Lecture Notes in Computer Science, Springer-Verlag, to appear.

M. E. SZABO (1981), *Variable truth*, to appear.

INFINITESIMAL ANALYSIS OF ℓ^∞ IN ITS MACKEY TOPOLOGY

K. D. Stroyan
Mathematical Sciences Division
The University of Iowa
Iowa City, Iowa 52242

Abstract: This expository article explains why the Mackey topology should play an important role in the analysis of the space of bounded sequences, ℓ^∞. We defend this on pure and applied grounds. We also describe some new results from Robinson's Theory of Infinitesimals that aid in this analysis.

Contents:

1. Conservation of Topologies on ℓ^∞:

Rubel [1971] states the "principle of conservation of topologies" which we rephrase as follows: Each (interesting) space has only a few (interesting) topologies. His article illustrates the effect of this principle on the space of bounded holomorphic functions. We shall illustrate its effect on ℓ^∞ which has even fewer topologies.

Here are some definitions of topologies on ℓ^∞ that arise from functional analysis constructions. The point of this section is that on ℓ^∞ they all coincide. Section 3 gives an applied characterization of this same topology.

(1.1) MACKEY: The Mackey topology of ℓ^∞ is defined to be the finest locally convex topology on ℓ^∞ whose dual is ℓ^1. Since ℓ^1 is not reflexive, this is coarser than the (strong =) norm topology. The Mackey-Arens theorem says that MACKEY is given by uniform convergence of ℓ^∞-sequences acting as functionals on weak $\sigma(\ell^1,\ell^\infty)$-compact convex subsets of ℓ^1.

(1.2) NORM COMPACT: The topology we denote by NORM COMPACT is the topology of uniform convergence on norm compact convex subsets of ℓ^1.

(1.3) NORM NULL: The topology we denote by NORM NULL is the topology of uniform convergence on ℓ^1-norm null sequences, $\|a_m\|_1 \to 0$.

(1.4) BOUNDED WEAK-STAR: The bounded weak-star topology is defined to be the finest topology on ℓ^∞ that agrees with the weak-star, $\sigma(\ell^\infty,\ell^1)$-topology on norm bounded sets. A set U is open if and only if for each m, the set $U \cap \{x : \|x\|_\infty \le m\}$ is $\sigma(\ell^\infty,\ell^1)$-open in $\|x\|_\infty \le m$.

(1.5) STRICT: Buck introduced the strict topology on the space of bounded continuous functions of a noncompact space. We may view $\ell^\infty(\mathbb{N})$ as the bounded continuous functions on $\mathbb{N}$. Seminorms for the strict topology are given by

$$|x|_c = \sup[\,|x_m c_m| : m \in \mathbb{N}]$$

for all null sequences, $c_m \longrightarrow 0$.

(1.6) BOUNDED POINTWISE: The notion of bounded pointwise convergence of a sequence of functions plays a role in classical analysis. Specialized to ℓ^∞ this would say, for $\{x^k\} \subseteq \ell^\infty$, $x^k \xrightarrow{B} y$ if and only if $\|x^k - y\|_\infty$ is uniformly bounded in k and for each fixed m, $\lim_{k\to\infty} x^k_m = y_m$. Results in Rubel and Ryff [1970] show that this sequential notion is compatible with a topology in the sense that there is a finest topology with these convergent sequences. We refer to this topology as BOUNDED POINTWISE.

(1.7) THEOREM:

All the topologies (1.1)-(1.6) _coincide and make_ (ℓ^∞, MACKEY) _a topological linear algebra under pointwise multiplication and addition._

All of these coincidences are known results and, moreover, one can give simple proofs on ℓ^∞. We will point out the main references and also show what role a simple constrained infinitesimal relation on $^*\ell^\infty$ can play in understanding part of this.

First, it is easy to see that NORM NULL $\subseteq$ NORM COMPACT. On ℓ^1 weak and norm compactness coincide (Dunford & Schwartz [1958, p. 339)], so NORM COMPACT = MACKEY. H. H. Schaeffer [1966, p. 150] shows that NORM COMPACT = BOUNDED WEAK-STAR in general and Dunford & Schwartz [1958, p. 427] shows that BOUNDED WEAK-STAR = NORM NULL in general. So far we see that (1.1) = (1.2) = (1.3) = (1.4).

STRICT = MACKEY is a result of Conway [1967], but this can also be seen as follows. We define a nontopological constrained infinitesimal relation as follows.

(1.8) DEFINITION:

For $x,y \in {}^*\ell^\infty$, _we write_ $x \overset{M}{\approx} y$ _if and only if_

(a) $\|x-y\|_\infty \in \mathcal{O}$ <u>is finite or</u> $(x_n - y_n)$ <u>is finite for all</u> $n \in {}^*\mathbb{N}$,

<u>and</u>

(b) $x_m \approx y_m$ <u>for all finite</u> $m \in {}^\sigma\mathbb{N}$.

Stroyan [1974] shows that a standard seminorn, p, is STRICTLY-continuous if and only if $x \overset{M}{\approx} 0$ implies $p(x) \approx 0$. If we let ℓ^∞ be the dual of $\ell^1[(\frac{1}{m})^2] = L^1(\mu)$, where $\mu(S) = \Sigma[(\frac{1}{m})^2 : m \in S]$, then Stroyan [1973] shows that a standard seminorm, p, is MACKEY$(\ell^\infty, \ell^1[(\frac{1}{m})^2])$-continuous if and only if $x \overset{M}{\approx} 0$ implies $p(x) \approx 0$. The reason is that the only subsets of ${}^*\mathbb{N}$ with infinitesimal ${}^*\mu$-measure are subsets of the infinite $*$-natural numbers. It is easy to see that MACKEY(ℓ^∞, ℓ^1) = MACKEY$(\ell^\infty, \ell^1[(\frac{1}{m})^2])$, thus STRICT = MACKEY.

Finally, what about BOUNDED POINTWISE. Stroyan [1974] shows that a sequence $x^k \xrightarrow{B} y$ if and only if $x^n \overset{M}{\approx} y$ for every infinite $n \in {}^*\mathbb{N}$. Thus MACKEY $\subseteq$ BOUNDED POINTWISE.

Rubel and Ryff [1980, p. 172] show that BOUNDED WEAK-STAR is the finest topology on the dual of a separable (B)-space which has the same convergent <u>sequences</u> as the weak-star topology. Hence, we only need show that $x^k \xrightarrow{B} y$ if and only if $x^k \xrightarrow{\text{wk-star}} y$. The uniform boundedness principle assures us that a weak-star convergent <u>sequence</u> will be uniformly bounded. Pointwise convergence comes simply from delta sequences in ℓ^1. The converse convergence implication is easy, so we see that all the above topologies coincide.

Continuity of multiplication $(x,y) \longrightarrow xy : (xy)_m = x_m \cdot y_m$ is easy to prove with the help of ordinary uniform-topological infinitesimals (in the unconstrained case of Stroyan & Luxemburg [1976, Ch. 10]) for the STRICT = MACKEY-topology, see Stroyan [1981.a]. Theorem (1.7) has the effect that we may characterize the Mackey infinitesimal relation, $\overset{\mathfrak{M}}{\approx}$, many ways, for example, let $\text{cpt}({}^*\ell^1) = \bigcup\{{}^*K : K \subseteq \ell^1$ is compact, convex$\}$. This is the same for norm-compact as for weak-compact.

(1.9) ORDINARY MACKEY INFINITESIMALS:

<u>For</u> $x,y \in {}^*\ell^\infty$, <u>the following are equivalent</u>. <u>When these hold we write</u> $x \overset{\mathfrak{M}}{\approx} y$.

(a) <u>For every</u> $a \in \text{cpt}({}^*\ell^\infty)$, <u>we have</u> $\sum_{m=1}^{\infty} a_m x_m \approx \sum_{m=1}^{\infty} a_m y_m$.

(b) <u>For every standard null sequence</u> $c \in c_0$, $c_m \to 0$, <u>we have</u> $x_n c_n \approx y_n c_n$ <u>for all</u> $n \in {}^*\mathbb{N}$.

Stroyan [1973 or 1974] shows that $x \overset{M}{\approx} y$ implies $x \overset{\mathfrak{M}}{\approx} y$. The converse implication fails in general, i.e., $x \overset{\mathfrak{M}}{\approx} y \not\Rightarrow x \overset{M}{\approx} y$.

2. <u>$(\ell^\infty$,MACKEY) is an (HM)-space</u>:

Henson and Moore [1973] introduced an interesting new class of topological vector spaces related to, but different from the nuclear, Schwartz and Montel spaces. They were interested in the question of when the infinitesimal hull of a space is invariant under changes in the nonstandard polyenlargement used to form finite mod infinitesimal elements. This happens if and only if the hull of the space is its standard completion in any one nonstandard polyenlargement. For example, note that the hull of a (B)-space is always a larger (B)-space. Schwartz spaces are (HM)-spaces, but there are non-Schwartz (HM)-spaces. A complete metrizable space is (HM) if and only if it is Montel, i.e., (HM) $\cap$ (F) = (FM). Stroyan [1978] showed that calculus is especially nice on (HM)-spaces. Bellenot [1976] and others have studied this class.

Stroyan [1981.a] observed that $(\ell^\infty$,MACKEY) is a "natural" example of a non-Montel (HM)-space. This could have been proved with Henson and Moore's [1973] methods, but it was not observed at that time. Moore has since shown that it is a Schwartz space.

(2.1) THEOREM:

$(\ell^\infty$,MACKEY) <u>is</u> <u>a</u> <u>complete</u> (HM)-<u>space</u>, <u>but</u> <u>it</u> <u>is</u> <u>nonmetrizable</u>, <u>nonbarrelled</u>, <u>non-Montel</u>.

The first part is proved in Stroyan [1981.a] by simply showing that pointwise standard part is an isomorphism: $(\ell^\infty,\text{MACKEY})^\wedge \xrightarrow{st} (\ell^\infty,\text{MACKEY})$. Specifically, if x is MACKEY-finite in $^*\ell^\infty$, that is if $p(x)$ is finite for each standard MACKEY-seminorm, then x is a sequence on $^*\mathbb{N}$. Let $\tilde{x}$ be the sequence given by taking

$$\tilde{x}_m = st(x_m),$$

when m is finite. Using STRICT = MACKEY, Stroyan [1981.a] calculates that $\tilde{x} \in \ell^\infty$ and $^*\tilde{x} \overset{\mathfrak{M}}{\approx} x$, see (1.9).
By (1.9) we only need to prove $x_n c_n \approx {}^*\tilde{x}_n c_n$, for all n when c is a standard null sequence.

Infinitesimal hulls are always complete (Stroyan & Luxemburg [1976, p. 273]), so ℓ^∞ is its own complete hull for every polyenlargement.

Rubel and Ryff [1970, p. 181] show that general BOUNDED WEAK-STAR topologies are never even first-countable, hence not metrizable.

A <u>barrell</u> is an absolutely convex closed absorbent set. A space is called <u>barrelled</u> when every barrel is a neighborhood of the origin. Barrelled spaces satisfy a Banach-Steinhaus theorem. Montel spaces are barrelled spaces with the property that bounded closed sets are compact.

The norm-ball $\{x \in \ell^\infty : \|x\|_\infty \le 1\}$ is MACKEY-closed because if $x \in {}^*\ell^\infty$ has $\|x\|_\infty \le 1$, then $\tilde{x} : \tilde{x}_m = st(x_m)$ has $\|\tilde{x}\| \le 1$. The ball is not a MACKEY-neighborhood of zero. In fact, since NORM $\neq$ MACKEY, there must be ordinary MACKEY-infinitesimals, $x \approx 0$, with infinite norm, see Stroyan & Luxemburg [1976, p. 274].

3. <u>Mackey Continuous Economic Agents Are Myopic</u>:

In this section we offer a simplified explanation of a result of Brown and Lewis [1981] that relates short-sighted economic behavior to the Mackey topology on ℓ^∞. We wish to study bounded sequences of consumption, so x_m represents the amount of a commodity consumed at the discrete time m. The set of admissible consumption sequences is the positive part of ℓ^∞,

$$\ell^\infty_+ = \{x \in \ell^\infty : x_m \ge 0 \text{ for all } m\}.$$

A functional $u : \ell^\infty \to \mathbb{R}_+$ is called a <u>utility function</u>. (Usually u has additional properties such as "more is better," but we don't need that. For technical reasons, we shall assume all our utility functions are defined on all ℓ^∞ not just ℓ^∞_+.) We may think of u as measuring which consumption sequences are preferred to others by saying the agent with utility u prefers x to y if $u(x) > u(y)$.

We say that a utility function is <u>myopic</u> provided that for all $x,y,z \in \ell^\infty$, if $u(x) > u(y)$, then $u(x) > (y+\bar{z}^m)$ for all sufficiently large m, where $\bar{z}^m_k = 0$ for $k \le m$ and $\bar{z}^m_k = z_k$ for $k > m$. This is one way to say that u is indifferent to any bounded increase in consumption if it takes place too far in the future. The method of constrained infinitesimals on ℓ^∞ of section 4 and (1.8) gives another intuitive way to say the same thing: <u>arbitrarily small changes in the finite future are preferred to any bounded change in the infinite future</u>. (Even a million dollars for every period after n... but why not if you're dead?)

MACKEY is the finest myopic linear topology, so it carries 'all' the myopic utility functions.

(3.1) THEOREM:

Every MACKEY-continuous $u : \ell_\infty \longrightarrow \mathbb{R}_+$ is myopic. Moreover, if τ is a locally convex Hausdorff topology on ℓ^∞ with the property that every τ-continuous functional is myopic, then τ is coarser than MACKEY.

The first part of the theorem follows from (1.7). Since BOUNDED POINTWISE = MACKEY and since $y + \bar{z}^m \xrightarrow{B} y$, continuity of u implies that $u(x) > \lim_m u(y+\bar{z}^m)$.

Instead of re-proving Brown=Lewis' theorem as it is stated, we will show that a similar formulation of myopia would yield their result quite easily. Let us say that a standard utility $u : \ell^\infty \longrightarrow \mathbb{R}_+$ is finitely foresighted if whenever $w \overset{M}{\approx} x \in \ell^\infty$, then $u(w) \approx u(x)$. [The relation $w \overset{M}{\approx} x$ is defined in (1.8).] We claim that this is a reasonable substitute for myopia on basic intuitive grounds because the only measurable changes in w over x occur at infinite times. Section 4 shows that it is technically equivalent in one sense (linearity). It is certainly easy to show that finite foresight implies myopia because $(y + \bar{z}^n) \overset{M}{\approx} y$ whenever n is infinite, so $u(y) \approx u(y + \bar{z}^n) < \frac{1}{2}[u(x) + u(y)]$ for all infinite n. The internal set $\{n \in {}^*\mathbb{N} : u(y+\bar{z}^n) < \frac{1}{2}[u(x) + u(y)]\}$ must contain a finite n, because the set of infinite *-natural numbers is external. Thus finitely foresighted implies myopic.

If τ is a locally convex topology such that all its utilities have finite foresight, then τ is coarser than MACKEY. The proof can be based on either Stroyan [1973] or [1974], since STRICT = MACKEY = MACKEY$(\ell^\infty, \ell^1[(\frac{1}{m})^2])$. In either case a τ-seminorm is MACKEY continuous because those references show that a seminorm p is MACKEY-continuous if and only if $x \overset{M}{\approx} 0$ implies $p(x) \approx 0$. In section 4 we will show that this continuity criterion holds for all functionals. Probably this is only because ℓ^∞ is special, that is, all the spaces in Stroyan [1973-4] probably don't have this continuity criterion. The result of section 4 does yield a stronger conclusion than Brown-Lewis's for the price of the stronger assumption.

(3.2) THEOREM:

If τ is any (Tychonoff) topology on ℓ^∞ such that all its utilities have finite foresight, then τ is coarser than MACKEY.

We do not need to assume that τ is linear. This result does not appear elsewhere, but it is a minor observation.

Stroyan [1981.b] gives explicit types of examples of myopic utility functions by showing that various kinds of calculations are finitely foresighted. Folloiwng are two examples of finitely foresighted functions. They are not economically interesting, but the first illustrates how one might obtain "stationary" utility functions.

(3.3) EXAMPLE:

Define $f : \ell^\infty \longrightarrow \mathbb{R}_+$ by

$$f(x) = \cfrac{1}{1 + \cfrac{x_1^2}{1 + \cfrac{1}{1 + \cfrac{x_2^2}{1 + \cfrac{1}{1 + \cfrac{x_3^2}{1 + \ddots}}}}}}$$

That is, let $V(x,u) = \dfrac{1}{1 + \dfrac{x^2}{1+|u|}}$ and take the recursive limit

$V(x_1,u);V(x_1,V(x_2,u));V(x_1,V(x_2,V(x_3,u)));\cdots$. This function is stationary in the sense that $V(x_1,f(x_2,x_3,\cdots)) = f(x_1,x_2,\cdots)$. Theorem 7 in Stroyan [1981.b] shows that f is Mackey continuous, by using the contracting property of V,

$$|V(x,u) - V(x,v)| \le \tfrac{1}{2}|u - v|$$

and the constrained continuity condition of section 4.

(3.4) EXAMPLE:

Define $u : \ell^\infty \longrightarrow \ell^\infty$ by

$$u(x) : [u(x)]_m = \cos(x_m)e^{x_m}.$$

This sends ℓ^∞ to ℓ^∞ clearly, but the rapid growth of e^{x_m} causes some difficulty in a classical check of MACKEY-continuity. In fact, even using ordinary MACKEY-infinitesimals we have some problems because we know there are $x \overset{M}{\approx} 0$ with x_n infinite when $n \in {}^*\mathbb{N}$ is infinite. Naturally, e^{x_m} is 'more infinite'...

It is easy to check u for finite foresight. If $x \in \ell^\infty$ is standard and $w \overset{M}{\approx} x$, then $\|w\|_\infty \in \mathcal{O}$, so $u(w)_n = \cos(w_n)e^{w_n} \in \mathcal{O}$.

When m is finite, ordinary continuity says, $\cos(w_m)e^{w_m} \approx \cos(x_m)e^{x_m}$, hence $u(x) \overset{M}{\approx} u(w)$.

In section 4 we say why this proves that u is MACKEY-continuous. In section 5 we show how the constrained calculation helps differentiate u with respect to MACKEY.

4. An Infinitesimal Continuity Criterion:

Suppose (X,τ) is an arbitrary (Tychonoff) topological space and $x \overset{\tau}{\approx} y$ denotes Wattenberg's topological infinitesimals of ${}^*X$, see Stroyan & Luxemburg, p. 222. If x is standard, $y \overset{\tau}{\approx} x$ if and only if $st(x) = st(y)$. To check MACKEY-continuity of $f : \ell^\infty \to X$ with ordinary infinitesimals, we would need to show that if $y \overset{\mathfrak{M}}{\approx} x \in \ell^\infty$, then $f(y) \overset{\tau}{\approx} f(x) \in X$. (See (1.9) and (1.8).) While not every $y \overset{\mathfrak{M}}{\approx} x$ satisfies $y \overset{M}{\approx} x$, it is a useful fact that these constrained infinitesimals suffice to check continuity.

(4.1) CONSTRAINED CONTINUITY LEMMA:

Let $f : \ell^\infty \to X$ be a standard function, for X as above. Then f is MACKEY-continuous if and only if whenever $y \overset{M}{\approx} x \in \ell^\infty$, then $f(y) \overset{\tau}{\approx} f(x) \in X$.

The proof of this result will appear in a later article. Try to verify MACKEY-continuity of EXAMPLES (3.3 & 4) without (4.1).

5. Mackey Differential Calculus:

Denote the MACKEY-finite points of ${}^*\ell^\infty$ by

$$\mathrm{FIN}({}^*\ell^\infty) = \{x \in \ell^\infty : p(x) \in \mathcal{O} \text{ for every standard MACKEY-seminorm}\}.$$

A standard function $u : \ell^\infty \xrightarrow{\tau} \ell^\infty$ is MACKEY-differentiable at x if there is a standard MACKEY-continuous linear transformation $D_xu : \ell^\infty \to \ell^\infty$ such that for every finite $z \in \mathrm{FIN}({}^*\ell^\infty)$ and every infinitesimal scalar $\delta \approx 0$,

$$u(x+\delta z) = u(x) + D_xu * \delta z + \delta\eta,$$

where $\eta \overset{\mathfrak{M}}{\approx} 0$. Consider the case of u in EXAMPLE (3.4) above. If anything is to be a derivative, certainly each coordinate must differentiate normally, so we have the coordinatewise definition for D_xu:

$$[D_xu * \delta z]_m = [(\cos(x_m) - \sin(x_m))e^{x_m}]\delta z_m.$$

First, we may use (4.1) to see that $x \longmapsto (\cos(x_m) - \sin(x_m))e^{x_m}$ is a MACKEY-continuous function. Next, we may use (2.1), the fact that $(\ell^{\infty},\text{MACKEY})$ is an (HM)-space, to see that $(\cos(x_m)-\sin(x_m))e^{x_m}$ is MACKEY-finite whenever x is. Since $(\ell^{\infty},\text{MACKEY})$ is a topological algebra, $D_x u$ is continuous. Now, it remains to show the "small oh" formula above assuring that $D_x u$ is the total derivative. This follows from Stroyan [1981.a, Theorem 18], which we state below as (5.1) in a specialized setting. We shall not repeat the proof here, but rather show how (2.1), (4.1) and the algebra property of ℓ^{∞} work together with (5.1).

(5.1) THEOREM:

Let $\{v_m(r) : m \in \mathbb{N}\}$ be a standard family of functions defined on the whole line, $v_m(r) : (-\infty,\infty) \longmapsto (-\infty,\infty)$. Suppose that there is a standard function $e(r,\theta)$ such that

$$|v_m(r+\Delta) - v_m(r) - v_m'(r)\Delta| \leq \Delta e(r,\Delta)$$

where $e(r,\Delta) \approx 0$ whenever r is finite and $\Delta \approx 0$ and such that

$$e(x,y) \overset{\mathfrak{M}}{\approx} 0 \quad \text{whenever} \quad x \in \text{FIN}({}^*\ell^{\infty}) \quad \text{and} \quad y \overset{\mathfrak{M}}{\approx} 0.$$

Moreover, suppose $v_m'(x_m) \in \text{FIN}({}^*\ell^{\infty})$ whenever $x \in \text{FIN}({}^*\ell^{\infty})$. Then the function $u(x) : [u(x)]_m = v_m(x_m)$ has MACKEY-derivative

$$D_x u * z : [D_x u * z]_m = v_m'(x_m) z_m.$$

To verify the remaining hypotheses of (5.1) we simply write the scalar Taylor formula for each coordinate obtaining one bound $e(r,\theta) = \theta e^{|r|+|\theta|}$, that is,

$$|\cos(x_m+\Delta_m)e^{x_m+\Delta_m} - \cos(x_m)e^{x_m} - (\cos(x_m)-\sin(x_m))e^{x_m}\Delta_m| \leq \Delta_m[\Delta_m e^{|x_m|+|\Delta_m|}].$$

Again (4.1) and (2.1) say that $e^{|x_m|+|y_m|}$ is MACKEY-finite whenever $x \in \text{FIN}({}^*\ell^{\infty})$ and $y \overset{\mathfrak{M}}{\approx} 0$. Since $(\ell^{\infty},\text{MACKEY})$ is an algebra, $e(x,y) \overset{\mathfrak{M}}{\approx} 0$ whenever $x \in \text{FIN}({}^*\ell^{\infty})$ and $y \overset{\mathfrak{M}}{\approx} 0$. Thus (5.1) says u is MACKEY-differentiable.

(5.3) EXAMPLE:

Define $w : \ell^{\infty} \longrightarrow \ell^{\infty}$ pointwise by

$$w(x) : [w(x)]_m = \frac{2}{\pi}\cos\left(\frac{\pi}{2} x_m^m\right)e^{x_m}.$$

Using (4.1) it is easy to verify that w is MACKEY-continuous. What about differentiability? First, the partial derivatives,

$$w'_m(x_m) = [\frac{2}{\pi}\cos(\frac{\pi}{2}x_m^m) - mx_m^{m-1}\sin(\frac{\pi}{2}x_m^m)]e^{x_m},$$

do not define a continuous linear transformation. That is, let $L(z):[L(z)]_m = w'_m(x_m) * z_m$. We may use (4.1), but more basically, if $x_m \equiv 1$, $z_m \equiv 1$, then

$$[L(z)]_m = -me \notin \ell^\infty.$$

Thus $w: \ell^\infty \longrightarrow \ell^\infty$ is continuous, but not differentiable except perhaps at a few strange points of ℓ^∞.

REFERENCES:

Steven F. Bellenot, On Nonstandard Hulls of Convex Spaces, Canad. J. Math. vol. XXVIII, nr. 1, pp. 141-147, [1976].

Donald J. Brown & Lucinda M. Lewis, Myopic Economic Agents, Econometrica, vol. 49, nr. 2, pp. 359-368, [1981].

J.B. Conway, The Strict Topology and Compactness in the Spaces of Measures, II, Trans. A.M.S., vol. 126, pp. 474-486, [1967].

N. Dunford & J. Schwartz, "Linear Operators, Part I," Interscience, New York [1958].

C. Ward Henson & L. C. Moore, Jr., Invariance of the Nonstandard Hulls of Locally Convex Spaces, Duke Math. J., vol. 40, pp. 193-205, [1973].

L.A. Rubel & J.V. Ryff, The Bounded Weak-Star Topology and the Bounded Analytic Functions, J. Funct'l. Anal., vol. 5, nr. 2, pp. 167-183, [1970].

L.A. Rubel, Bounded Convergence of Analytic Functions, Bull. A.M.S., vol. 77, pp. 13-24, [1971].

H.H. Schaefer, "Topologcial Vector Spaces," Macmillan, New York [1966].

K. D. Stroyan & W.A.J. Luxemburg, "Introduction to the Theory of Infinitesimals," Academic Press, New York, [1976].

K.D. Stroyan, A Characterization of the Mackey Uniformity $m(L^\infty, L^1)$ for Finite Measures, Pacific J. Math., vol. 49, nr. 1, pp. 223-228, [1973].

__________, A Nonstandard Characterization of Mixed Topologies, in vol. (the first), Victoria Symposium on Nonstandard Analysis, Hurd & Loeb (editors), Springer-Verlag Lecture Notes 369, New York [1974].

__________, Infinitesimal Calculus on Locally Convex Spaces 1: Fundamentals, Trans. A.M.S., vol. 240, pp. 363-383, [1978].

__________, Locally Convex Infinitesimal Calculus 2: Computations on Mackey (ℓ^∞), submitted, [1981.a].

__________, Myopic Utility Functions on Sequential Economies, submitted, [1981.b].

Vol. 844: Groupe de Brauer. Proceedings. Edited by M. Kervaire and M. Ojanguren. VII, 274 pages. 1981.

Vol. 845: A. Tannenbaum, Invariance and System Theory: Algebraic and Geometric Aspects. X, 161 pages. 1981.

Vol. 846: Ordinary and Partial Differential Equations, Proceedings. Edited by W. N. Everitt and B. D. Sleeman. XIV, 384 pages. 1981.

Vol. 847: U. Koschorke, Vector Fields and Other Vector Bundle Morphisms – A Singularity Approach. IV, 304 pages. 1981.

Vol. 848: Algebra, Carbondale 1980. Proceedings. Ed. by R. K. Amayo. VI, 298 pages. 1981.

Vol. 849: P. Major, Multiple Wiener-Itô Integrals. VII, 127 pages. 1981.

Vol. 850: Séminaire de Probabilités XV. 1979/80. Avec table générale des exposés de 1966/67 à 1978/79. Edited by J. Azéma and M. Yor. IV, 704 pages. 1981.

Vol. 851: Stochastic Integrals. Proceedings, 1980. Edited by D. Williams. IX, 540 pages. 1981.

Vol. 852: L. Schwartz, Geometry and Probability in Banach Spaces. X, 101 pages. 1981.

Vol. 853: N. Boboc, G. Bucur, A. Cornea, Order and Convexity in Potential Theory: H-Cones. IV, 286 pages. 1981.

Vol. 854: Algebraic K-Theory. Evanston 1980. Proceedings. Edited by E. M. Friedlander and M. R. Stein. V, 517 pages. 1981.

Vol. 855: Semigroups. Proceedings 1978. Edited by H. Jürgensen, M. Petrich and H. J. Weinert. V, 221 pages. 1981.

Vol. 856: R. Lascar, Propagation des Singularités des Solutions d'Equations Pseudo-Différentielles à Caractéristiques de Multiplicités Variables. VIII, 237 pages. 1981.

Vol. 857: M. Miyanishi. Non-complete Algebraic Surfaces. XVIII, 244 pages. 1981.

Vol. 858: E. A. Coddington, H. S. V. de Snoo: Regular Boundary Value Problems Associated with Pairs of Ordinary Differential Expressions. V, 225 pages. 1981.

Vol. 859: Logic Year 1979–80. Proceedings. Edited by M. Lerman, J. Schmerl and R. Soare. VIII, 326 pages. 1981.

Vol. 860: Probability in Banach Spaces III. Proceedings, 1980. Edited by A. Beck. VI, 329 pages. 1981.

Vol. 861: Analytical Methods in Probability Theory. Proceedings 1980. Edited by D. Dugué, E. Lukacs, V. K. Rohatgi. X, 183 pages. 1981.

Vol. 862: Algebraic Geometry. Proceedings 1980. Edited by A. Libgober and P. Wagreich. V, 281 pages. 1981.

Vol. 863: Processus Aléatoires à Deux Indices. Proceedings, 1980. Edited by H. Korezlioglu, G. Mazziotto and J. Szpirglas. V, 274 pages. 1981.

Vol. 864: Complex Analysis and Spectral Theory. Proceedings, 1979/80. Edited by V. P. Havin and N. K. Nikol'skii, VI, 480 pages. 1981.

Vol. 865: R. W. Bruggeman, Fourier Coefficients of Automorphic Forms. III, 201 pages. 1981.

Vol. 866: J.-M. Bismut, Mécanique Aléatoire. XVI, 563 pages. 1981.

Vol. 867: Séminaire d'Algèbre Paul Dubreil et Marie-Paule Malliavin. Proceedings, 1980. Edited by M.-P. Malliavin. V, 476 pages. 1981.

Vol. 868: Surfaces Algébriques. Proceedings 1976–78. Edited by J. Giraud, L. Illusie et M. Raynaud. V, 314 pages. 1981.

Vol. 869: A. V. Zelevinsky, Representations of Finite Classical Groups. IV, 184 pages. 1981.

Vol. 870: Shape Theory and Geometric Topology. Proceedings, 1981. Edited by S. Mardešić and J. Segal. V, 265 pages. 1981.

Vol. 871: Continuous Lattices. Proceedings, 1979. Edited by B. Banaschewski and R.-E. Hoffmann. X, 413 pages. 1981.

Vol. 872: Set Theory and Model Theory. Proceedings, 1979. Edited by R. B. Jensen and A. Prestel. V, 174 pages. 1981.

Vol. 873: Constructive Mathematics, Proceedings, 1980. Edited by F. Richman. VII, 347 pages. 1981.

Vol. 874: Abelian Group Theory. Proceedings, 1981. Edited by R. Göbel and E. Walker. XXI, 447 pages. 1981.

Vol. 875: H. Zieschang, Finite Groups of Mapping Classes of Surfaces. VIII, 340 pages. 1981.

Vol. 876: J. P. Bickel, N. El Karoui and M. Yor. Ecole d'Eté de Probabilités de Saint-Flour IX – 1979. Edited by P. L. Hennequin. XI, 280 pages. 1981.

Vol. 877: J. Erven, B.-J. Falkowski, Low Order Cohomology and Applications. VI, 126 pages. 1981.

Vol. 878: Numerical Solution of Nonlinear Equations. Proceedings, 1980. Edited by E. L. Allgower, K. Glashoff, and H.-O. Peitgen. XIV, 440 pages. 1981.

Vol. 879: V. V. Sazonov, Normal Approximation – Some Recent Advances. VII, 105 pages. 1981.

Vol. 880: Non Commutative Harmonic Analysis and Lie Groups. Proceedings, 1980. Edited by J. Carmona and M. Vergne. IV, 553 pages. 1981.

Vol. 881: R. Lutz, M. Goze, Nonstandard Analysis. XIV, 261 pages. 1981.

Vol. 882: Integral Representations and Applications. Proceedings, 1980. Edited by K. Roggenkamp. XII, 479 pages. 1981.

Vol. 883: Cylindric Set Algebras. By L. Henkin, J. D. Monk, A. Tarski, H. Andréka, and I. Németi. VII, 323 pages. 1981.

Vol. 884: Combinatorial Mathematics VIII. Proceedings, 1980. Edited by K. L. McAvaney. XIII, 359 pages. 1981.

Vol. 885: Combinatorics and Graph Theory. Edited by S. B. Rao. Proceedings, 1980. VII, 500 pages. 1981.

Vol. 886: Fixed Point Theory. Proceedings, 1980. Edited by E. Fadell and G. Fournier. XII, 511 pages. 1981.

Vol. 887: F. van Oystaeyen, A. Verschoren, Non-commutative Algebraic Geometry, VI, 404 pages. 1981.

Vol. 888: Padé Approximation and its Applications. Proceedings, 1980. Edited by M. G. de Bruin and H. van Rossum. VI, 383 pages. 1981.

Vol. 889: J. Bourgain, New Classes of $\mathcal{L}^p$-Spaces. V, 143 pages. 1981.

Vol. 890: Model Theory and Arithmetic. Proceedings, 1979/80. Edited by C. Berline, K. McAloon, and J.-P. Ressayre. VI, 306 pages. 1981.

Vol. 891: Logic Symposia, Hakone, 1979, 1980. Proceedings, 1979, 1980. Edited by G. H. Müller, G. Takeuti, and T. Tugué. XI, 394 pages. 1981.

Vol. 892: H. Cajar, Billingsley Dimension in Probability Spaces. III, 106 pages. 1981.

Vol. 893: Geometries and Groups. Proceedings. Edited by M. Aigner and D. Jungnickel. X, 250 pages. 1981.

Vol. 894: Geometry Symposium. Utrecht 1980, Proceedings. Edited by E. Looijenga, D. Siersma, and F. Takens. V, 153 pages. 1981.

Vol. 895: J.A. Hillman, Alexander Ideals of Links. V, 178 pages. 1981.

Vol. 896: B. Angéniol, Familles de Cycles Algébriques – Schéma de Chow. VI, 140 pages. 1981.

Vol. 897: W. Buchholz, S. Feferman, W. Pohlers, W. Sieg, Iterated Inductive Definitions and Subsystems of Analysis: Recent Proof-Theoretical Studies. V, 383 pages. 1981.

Vol. 898: Dynamical Systems and Turbulence, Warwick, 1980. Proceedings. Edited by D. Rand and L.-S. Young. VI, 390 pages. 1981.

Vol. 899: Analytic Number Theory. Proceedings, 1980. Edited by M.I. Knopp. X, 478 pages. 1981.

Vol. 900: P. Deligne, J. S. Milne, A. Ogus, and K.-Y. Shih, Hodge Cycles, Motives, and Shimura Varieties. V, 414 pages. 1982.

Vol. 901: Séminaire Bourbaki vol. 1980/81 Exposés 561–578. III, 299 pages. 1981.

Vol. 902: F. Dumortier, P.R. Rodrigues, and R. Roussarie, Germs of Diffeomorphisms in the Plane. IV, 197 pages. 1981.

Vol. 903: Representations of Algebras. Proceedings, 1980. Edited by M. Auslander and E. Lluis. XV, 371 pages. 1981.

Vol. 904: K. Donner, Extension of Positive Operators and Korovkin Theorems. XII, 182 pages. 1982.

Vol. 905: Differential Geometric Methods in Mathematical Physics. Proceedings, 1980. Edited by H.-D. Doebner, S.J. Andersson, and H.R. Petry. VI, 309 pages. 1982.

Vol. 906: Séminaire de Théorie du Potentiel, Paris, No. 6. Proceedings. Edité par F. Hirsch et G. Mokobodzki. IV, 328 pages. 1982.

Vol. 907: P. Schenzel, Dualisierende Komplexe in der lokalen Algebra und Buchsbaum-Ringe. VII, 161 Seiten. 1982.

Vol. 908: Harmonic Analysis. Proceedings, 1981. Edited by F. Ricci and G. Weiss. V, 325 pages. 1982.

Vol. 909: Numerical Analysis. Proceedings, 1981. Edited by J.P. Hennart. VII, 247 pages. 1982.

Vol. 910: S.S. Abhyankar, Weighted Expansions for Canonical Desingularization. VII, 236 pages. 1982.

Vol. 911: O.G. Jørsboe, L. Mejlbro, The Carleson-Hunt Theorem on Fourier Series. IV, 123 pages. 1982.

Vol. 912: Numerical Analysis. Proceedings, 1981. Edited by G. A. Watson. XIII, 245 pages. 1982.

Vol. 913: O. Tammi, Extremum Problems for Bounded Univalent Functions II. VI, 168 pages. 1982.

Vol. 914: M. L. Warshauer, The Witt Group of Degree k Maps and Asymmetric Inner Product Spaces. IV, 269 pages. 1982.

Vol. 915: Categorical Aspects of Topology and Analysis. Proceedings, 1981. Edited by B. Banaschewski. XI, 385 pages. 1982.

Vol. 916: K.-U. Grusa, Zweidimensionale, interpolierende Lg-Splines und ihre Anwendungen. VIII, 238 Seiten. 1982.

Vol. 917: Brauer Groups in Ring Theory and Algebraic Geometry. Proceedings, 1981. Edited by F. van Oystaeyen and A. Verschoren. VIII, 300 pages. 1982.

Vol. 918: Z. Semadeni, Schauder Bases in Banach Spaces of Continuous Functions. V, 136 pages. 1982.

Vol. 919: Séminaire Pierre Lelong – Henri Skoda (Analyse) Années 1980/81 et Colloque de Wimereux, Mai 1981. Proceedings. Edité par P. Lelong et H. Skoda. VII, 383 pages. 1982.

Vol. 920: Séminaire de Probabilités XVI, 1980/81. Proceedings. Edité par J. Azéma et M. Yor. V, 622 pages. 1982.

Vol. 921: Séminaire de Probabilités XVI, 1980/81. Supplément: Géométrie Différentielle Stochastique. Proceedings. Edité par J. Azéma et M. Yor. III, 285 pages. 1982.

Vol. 922: B. Dacorogna, Weak Continuity and Weak Lower Semicontinuity of Non-Linear Functionals. V, 120 pages. 1982.

Vol. 923: Functional Analysis in Markov Processes. Proceedings, 1981. Edited by M. Fukushima. V, 307 pages. 1982.

Vol. 924: Séminaire d'Algèbre Paul Dubreil et Marie-Paule Malliavin. Proceedings, 1981. Edité par M.-P. Malliavin. V, 461 pages. 1982.

Vol. 925: The Riemann Problem, Complete Integrability and Arithmetic Applications. Proceedings, 1979-1980. Edited by D. Chudnovsky and G. Chudnovsky. VI, 373 pages. 1982.

Vol. 926: Geometric Techniques in Gauge Theories. Proceedings, 1981. Edited by R. Martini and E.M.de Jager. IX, 219 pages. 1982.

Vol. 927: Y. Z. Flicker, The Trace Formula and Base Change for GL (3). XII, 204 pages. 1982.

Vol. 928: Probability Measures on Groups. Proceedings 1981. Edited by H. Heyer. X, 477 pages. 1982.

Vol. 929: Ecole d'Eté de Probabilités de Saint-Flour X – 1980. Proceedings, 1980. Edited by P.L. Hennequin. X, 313 pages. 1982.

Vol. 930: P. Berthelot, L. Breen, et W. Messing, Théorie de Dieudonné Cristalline II. XI, 261 pages. 1982.

Vol. 931: D.M. Arnold, Finite Rank Torsion Free Abelian Groups and Rings. VII, 191 pages. 1982.

Vol. 932: Analytic Theory of Continued Fractions. Proceedings, 1981. Edited by W.B. Jones, W.J. Thron, and H. Waadeland. VI, 240 pages. 1982.

Vol. 933: Lie Algebras and Related Topics. Proceedings, 1981. Edited by D. Winter. VI, 236 pages. 1982.

Vol. 934: M. Sakai, Quadrature Domains. IV, 133 pages. 1982.

Vol. 935: R. Sot, Simple Morphisms in Algebraic Geometry. IV, 146 pages. 1982.

Vol. 936: S.M. Khaleelulla, Counterexamples in Topological Vector Spaces. XXI, 179 pages. 1982.

Vol. 937: E. Combet, Intégrales Exponentielles. VIII, 114 pages. 1982.

Vol. 938: Number Theory. Proceedings, 1981. Edited by K. Alladi. IX, 177 pages. 1982.

Vol. 939: Martingale Theory in Harmonic Analysis and Banach Spaces. Proceedings, 1981. Edited by J.-A. Chao and W.A. Woyczyński. VIII, 225 pages. 1982.

Vol. 940: S. Shelah, Proper Forcing. XXIX, 496 pages. 1982.

Vol. 941: A. Legrand, Homotopie des Espaces de Sections. VII, 132 pages. 1982.

Vol. 942: Theory and Applications of Singular Perturbations. Proceedings, 1981. Edited by W. Eckhaus and E.M. de Jager. V, 363 pages. 1982.

Vol. 943: V. Ancona, G. Tomassini, Modifications Analytiques. IV, 120 pages. 1982.

Vol. 944: Representations of Algebras. Workshop Proceedings 1980. Edited by M. Auslander and E. Lluis. V, 258 pages. 1982.

Vol. 945: Measure Theory. Oberwolfach 1981, Proceedings. Edited by D. Kölzow and D. Maharam-Stone. XV, 431 pages. 1982.

Vol. 946: N. Spaltenstein, Classes Unipotentes et Sous-groupes de Borel. IX, 259 pages. 1982.

Vol. 947: Algebraic Threefolds. Proceedings, 1981. Edited by A. Conte. VII, 315 pages. 1982.

Vol. 948: Functional Analysis. Proceedings, 1981. Edited by D. Butković, H. Kraljević, and S. Kurepa. X, 239 pages. 1982.

Vol. 949: Harmonic Maps. Proceedings, 1980. Edited by R.J. Knil M. Kalka and H.C.J. Sealey. V, 158 pages. 1982.

Vol. 950: Complex Analysis. Proceedings, 1980. Edited by J. Eells IV, 428 pages. 1982.

Vol. 951: Advances in Non-Commutative Ring Theory. Proceeding 1981. Edited by P.J. Fleury. V, 142 pages. 1982.

Vol. 952: Combinatorial Mathematics IX. Proceedings, 1981. Edite by E. Billington, S. Oates-Williams, and A.P. Street. XI, 443 page 1982.